Computational Thinking for Problem Solving and Managerial Mindset Training

Luisa Dall'Acqua
University of Bologna, Italy & LS TCO, Italy

A volume in the Advances in Business Strategy and Competitive Advantage (ABSCA) Book Series

IGI Global
PUBLISHER of TIMELY KNOWLEDGE

Published in the United States of America by
IGI Global
Business Science Reference (an imprint of IGI Global)
701 E. Chocolate Avenue
Hershey PA, USA 17033
Tel: 717-533-8845
Fax: 717-533-8661
E-mail: cust@igi-global.com
Web site: http://www.igi-global.com

Library of Congress Cataloging-in-Publication Data

Names: Dall'Acqua, Luisa, 1962- editor.
Title: Computational thinking for problem solving and managerial mindset training / Luisa Dall'Acqua, editor.
Description: Hershey, PA : Business Science Reference, 2022. | Includes bibliographical references and index. | Summary: "This book expores computational thinking (CT) which involves a set of skills related to the foundations of computer science as a scientific discipline that can be applied to our increasingly digital society, where software technological tools permeate daily life and, consequently, change the management of reality, and mastering of trasversal skills crucial to have success"-- Provided by publisher.
Identifiers: LCCN 2021023064 (print) | LCCN 2021023065 (ebook) | ISBN 9781799871262 (hardcover) | ISBN 9781799871279 (paperback) | ISBN 9781799871286 (ebook)
Subjects: LCSH: Information technology--Management. | Problem-solving--Computer-assisted instruction. | Management--Computer-assisted instruction. | Computational learning theory.
Classification: LCC HD30.2 .C6396 2022 (print) | LCC HD30.2 (ebook) | DDC 658.4/038--dc23
LC record available at https://lccn.loc.gov/2021023064
LC ebook record available at https://lccn.loc.gov/2021023065

This book is published in the IGI Global book series Advances in Business Strategy and Competitive Advantage (ABSCA) (ISSN: 2327-3429; eISSN: 2327-3437)

British Cataloguing in Publication Data
A Cataloguing in Publication record for this book is available from the British Library.

Advances in Business Strategy and Competitive Advantage (ABSCA) Book Series

ISSN:2327-3429
EISSN:2327-3437

Editor-in-Chief: Patricia Ordóñez de Pablos, Universidad de Oviedo, Spain

MISSION

Business entities are constantly seeking new ways through which to gain advantage over their competitors and strengthen their position within the business environment. With competition at an all-time high due to technological advancements allowing for competition on a global scale, firms continue to seek new ways through which to improve and strengthen their business processes, procedures, and profitability.

The **Advances in Business Strategy and Competitive Advantage (ABSCA) Book Series** is a timely series responding to the high demand for state-of-the-art research on how business strategies are created, implemented and re-designed to meet the demands of globalized competitive markets. With a focus on local and global challenges, business opportunities and the needs of society, the **ABSCA** encourages scientific discourse on doing business and managing information technologies for the creation of sustainable competitive advantage.

COVERAGE

- Small and Medium Enterprises
- Strategy Performance Management
- Competitive Strategy
- Adaptive Enterprise
- Cost Leadership Strategy
- Tacit Knowledge
- Value Chain
- Co-operative Strategies
- Core Competencies
- Business Models

IGI Global is currently accepting manuscripts for publication within this series. To submit a proposal for a volume in this series, please contact our Acquisition Editors at Acquisitions@igi-global.com or visit: http://www.igi-global.com/publish/.

The Advances in Business Strategy and Competitive Advantage (ABSCA) Book Series (ISSN 2327-3429) is published by IGI Global, 701 E. Chocolate Avenue, Hershey, PA 17033-1240, USA, www.igi-global.com. This series is composed of titles available for purchase individually; each title is edited to be contextually exclusive from any other title within the series. For pricing and ordering information please visit http://www.igi-global.com/book-series/advances-business-strategy-competitive-advantage/73672. Postmaster: Send all address changes to above address.

Titles in this Series

For a list of additional titles in this series, please visit:
http://www.igi-global.com/book-series/advances-business-strategy-competitive-advantage/73672

Handbook of Research on Disruptive Innovation and Digital Transformation in Asia
Patricia Ordóñez de Pablos (The University of Oviedo, Spain) Xi Zhang (Tianjin University, China) and Mohammad Nabil Almunawar (School of Business and Economics, University of Brunei Darussalam, Brunei)
Business Science Reference • ©2021 • 661pp • H/C (ISBN: 9781799864776) • US $295.00

Disruptive Technology and Digital Transformation for Business and Government
Kamaljeet Sandhu (University of New England, Australia)
Business Science Reference • ©2021 • 410pp • H/C (ISBN: 9781799885832) • US $225.00

Handbook of Research on Entrepreneurship, Innovation, Sustainability, and ICTs in the Post-COVID-19 Era
Luisa Cagica Carvalho (Polytechnic Institute of Setubal, Portugal & CEFAGE, University of Évora, Portugal) Leonilde Reis (Polytechnic Institute of Setubal, Portugal) and Clara Silveira (Polytechnic Institute of Guarda, Portugal)
Business Science Reference • ©2021 • 451pp • H/C (ISBN: 9781799867760) • US $295.00

Disruptive Technologies and Innovative Strategies for Organizational Success
Nilsa I. Elias (Western Governors University, USA) Terry M. Walker (University of Louisiana, USA) and Robin Walker (Metropolitan State University, USA)
Business Science Reference • ©2021 • 315pp • H/C (ISBN: 9781799837909) • US $225.00

Cases on Digital Strategies and Management Issues in Modern Organizations
José Duarte Santos (Instituto Superior Politecnico, Spain)
Business Science Reference • ©2021 • 300pp • H/C (ISBN: 9781799816300) • US $195.00

Advanced Perspectives on Global Industry Transitions and Business Opportunities
Fanny Saruchera (University of the Witwatersrand, South Africa)
Business Science Reference • ©2021 • 352pp • H/C (ISBN: 9781799843030) • US $225.00

For an entire list of titles in this series, please visit:
http://www.igi-global.com/book-series/advances-business-strategy-competitive-advantage/73672

IGI Global
PUBLISHER of TIMELY KNOWLEDGE

701 East Chocolate Avenue, Hershey, PA 17033, USA
Tel: 717-533-8845 x100 • Fax: 717-533-8661
E-Mail: cust@igi-global.com • www.igi-global.com

Table of Contents

Detailed Table of Contents

The chapter intends to be a theoretical contribution for developers in the field of artificial intelligence. It also means a practical guideline for leaders, as decision-makers, to manage tasks and optimize performance. The proposed approach interprets the fluid nature of the decision-making process looking at knowledge and knowledge activities as dynamic, adaptive, and self-regulative, based not only on well-known explicit curricular goals but also on unpredictable interactions and relationships between players. The knowledge process is emerging in human and biological, social, and cultural environments.

In critical activities and organizations, decision making in the face of complexity has been a growing normal. Complexity troubles humans due to cognitive limitations. Moreover, humans are merely able to understand cause-and-effect relationships that are close in time and space, not the paradigm of many complex socio-technical systems. Decision-making processes shall rely on models that help harness a problem´s associated complexity – among them the dynamics of supply chains. Models typically fall into two broad categories: mental and formal models. Supply chains are complex systems, which may exhibit complex behaviour patterns. Decisions and policies within organizational systems are the causes of many problems, among them undesirable oscillations and other problematic patterns of the parameters of interest. A system is a grouping of parts that work together for a purpose. Hence, the systems dynamics methodology is an adequate approach to deal with fuel supply chain management. A model was developed that helps manage marine gasoil supply chains in the context of the navy.

This chapter aims to investigate the effectiveness of managerial strategies to address sources of risk and strengthen global supply chain security, with specific attention to disasters deriving from the interaction between extreme natural phenomena and pre-existing socio-economic vulnerabilities. The object of the study is represented by the mutual learning opportunities that can arise from socially-oriented cross-sector collaborations between humanitarian agencies and enterprises. With the ultimate scope of grasping the economic impacts, the business perspective provides several insights to integrate a proactive approach into the corporate strategic planning to ensure a capacity to react with respect to external stimuli. In particular, an in-depth study was carried out dedicated to the regulatory standards that allow the development of this management system. Finally, the private sector's involvement in humanitarian programs is analyzed in order to highlight the benefits in terms of effectiveness of operational performances and the safety of communities' critical infrastructure.

Recent advancements in extended reality (XR) immersive technologies provide new tools for the development of novel and promising applications for business. Specifically, extended reality training applications are becoming popular in business due to their advantages of low cost, risk-free, data-oriented training. Extended reality training is the digital simulation of lifelike scenarios for training purposes using technologies such as virtual reality, augmented reality, mixed reality. Many applications are already available to train employees to develop specific technical skills, from maintenance to construction. The purpose of this chapter is to review the emerging XR applications developed for management training. Specifically, this chapter will focus on the training of some key skills in management such as leadership, problem solving, emotional intelligence, communication, and team working.

as being the result of changes in the way of thinking of new generations, often caused by the negative impact of technology – its source is usually in the process of teaching and growth itself. Computational thinking allows us to take complex problems, understand what the problem is, and develop solutions. We can present these solutions in a way that both computers and people can understand.

Intelligence is a field of activity that has a rich history, yet it still lacks clarity of definition and agreement. This is not to denigrate the performance of many of its practitioners; rather, these comments are designed to point up the fact that, despite the length and breadth of its historical practice, there is still much to do to explore the boundaries, opportunities, and limitations of the application of intelligence in the world of enforcement. The cultural roots of strategic intelligence is compelling and urgent not just for those involved with the professional intelligence community, but also for anyone seriously committed to interdisciplinary studies, cross-cultural understanding, and most importantly, to the development of a rigorous discipline of politics as well as cultural genetics. But strategic planning does not guarantee strategic thinking.

Last January, the General Secretariat of the Council of the European Union produced an estimate that, by the year 2035, the drones' sector will create up to 100,000 new jobs and, in addition, will have an economic impact of more than ten billion euros per year. This chapter takes its cue precisely from the central importance of unmanned aerial vehicles, commonly known as drones, and tries to sum up their future. Indeed, despite the enormous success of drones over the decades, many are the criticisms related to unmanned aircraft on board. The first section aims at outlining an historical framework, while the second section analyses the impressive role played by drones within the counterterrorism policy of the US presidency, with particular regard to Barack Obama's mandate. The third section applies a SWOT type analysis to UAVs, taking into account both their advantages and disadvantages. Finally, the fourth section investigates the human factor problem within UAVs, which was inspired by Dr. Giovanni Miranda's PhD thesis.

The term strategy model appears in a variety of contexts. Though the specifics of a strategy model vary greatly depending upon the nature of the model, the basic definition of such a model remains the same in all industries and sectors, such as in policy management. Basically, a strategy model constitutes a strategic plan, or model, designed to improve a process, and it needs a specific mindset by the developer. Organizations use strategy models to improve operations and meet their goals. The development of such a model requires identifying the key goals of an organization, identifying the key components or steps of the process leading from a starting point to the achievement of those goals and creating ways of maximizing the process for optimal output and the achievement of the stated goals. Strategy models are of varying degrees of complexity.

Preface

One of the fundamental characteristic properties of all living beings is that of being objects endowed with a project [...] a property to which we will give the name of teleonomy. [...] All the functional adaptations of living beings [...] carry out particular projects that can be considered as aspects or fragments of a single primitive project, that is, the conservation of the species. [...] All the structures, the performances, the activities that contribute to the success of the essential project will therefore be called 'teleonomic'. (Monod, 1970, pp. 22, 27)

At the level of computational theory, a mechanism is defined based on the purpose for which it is designed, in a sort of cognitive engine optimization. Although it is quite natural to attribute a purpose to an artificial mechanism, it is useful to say a few words about the legitimacy of such attribution to various functions, from biological to social or more generally managerial.

Once the purpose of a function has been established, it is necessary to define the resources it has. When designing a mechanism from scratch, it is possible to choose the machine that is best suited to carry out the project; but our aim is not to build a new mechanism but to manage an already given mechanism, and to do this we must determine the resources that it can dispose of if only to immediately eliminate implausible hypotheses. The logical sequence proper to the design of a system, be it artificial or natural, is therefore reversed in the study of a biological system: in this case, the physical machine is already given and it constitutes a constraint to which a computational model must be respected.

The cultural, social, and economic history of mankind is characterized by a succession of needs and problems that have stimulated the invention of operational and conceptual tools to facilitate their solution. The continuous presentation of new needs, an attempt to improve partial solutions to old problems, curiosity, and the disinterested search for knowledge then constituted the fundamental push for scientific, cultural, economic, and social progress. In an increasingly digital society, where software technological tools permeate daily life and, consequently, change the management of reality, mastering of transversal skills is crucial for success.

Let's take an anecdotal example. In the center of the city, there is a very convenient underground car park, because it is a stone's throw from your workplace. The problem is to orient yourself to understand where the lifts are, so much so that sometimes you can have the advantage of finding a place immediately and close, but the disadvantage of wasting time getting out on foot. Or that sometimes, when you get out with the car, you have to get out of one (allowed by the indicator arrows) which forces you to make several maneuvers, even reversing, to get out. This anecdote tells us some things. The first is that sometimes for some people and in particular situations it is difficult to orient themselves. But that's not the biggest problem. The second is that - when we make choices - we implicitly assess the costs and benefits. That parking is - in economic terms - relatively more expensive than other possible solutions, but it is very practical because it is close to the place where I have to go. Therefore a saving in terms of time and physical effort. But if I have to spend 5 minutes to find the exit, the time advantage decreases.

If a working "space" is poorly organized, the executive act can become more tiring and more expensive in terms of time. This is true for physical spaces, it is even more true for information domains.

This volume explores the potential of a method that is effective and explores how individuals can be trained into managerial mindsets through computational thinking and computer science. It explores how computer science can be used as a valid guideline to develop skills such as effective soft skills, communication skills, and collaboration.

Understanding how people see the world, how they represent knowledge, how they exchange it with the environment and with other people, and what their strengths and limitations are can help in the attempt to create valuable information systems. Good governance must necessarily go through obtaining understanding, a trained mindset, good data, and information. These will allow managers to know the past and the momentum of the business and, also with them, to predict, estimate and make the best-informed decisions. The greater the complexity of the business or tasks, the greater this need.

Knowledge implies a number of skills:

- the ability to recognize the salient aspects of a context (natural, social, cultural), to recognize patterns
- the possession of a repertoire of rules to make choices based on the motivations, the situation, the resources available
- Possession of a script repertoire of actions
- the ability to identify opportunities and affordances
- the ability to create models of the domain
- the ability to make predictions

- the ability to make choices and decisions based on the model
- the ability to plan activities, based on the model
- the ability to undertake the actions foreseen in the activity plan (execute the plan)
- the ability to monitor the process of the activity and actions
- the ability to evaluate the process and the results, also based on expectations

THE MENTALITY OF A PROJECT MANAGER

In this context, projects for the development of "intelligent" mindset training solutions, with huge impact and scope, undergo the need for continuous improvement and incremental evolution. Computational thinking, as well as a compared Agile method, by their nature and principles, are suitable to fulfill this need.

The term "intelligence" refers to the field of military science (Mettler, 2009). In fact, military intelligence is a discipline that applies data analysis methodologies to the data and information collected to provide guidance and direction in commanders' decisions (Austin, 1995). In a war context, the "commanders" of organizations, units, departments or services, need this "intelligence" to make the best decisions and deliver the best possible value to stakeholders. Thus, Business Intelligence (BI) is generally understood as the tool to transform organizations' existing data into knowledge to help them make informed decisions (Cao, 2013). Evelson (2008) defines BI as "a set of methodologies, processes, architectures, and technologies that transform raw data into meaningful and useful information used to enable more effective strategic, tactical, and operational insights and decision-making". In fact, BI can be described as "an umbrella term that combines architectures, tools, methodologies, databases and data warehouses, analytical tools, and applications." (El Morr, 2019a, p. 5-6).

BI is defined as a set of metrics to measure past performance and report a set of indicators that can guide decision-making, and Advanced Analytics as the analysis of trends, recognition of patterns and prescribe actions for better outcomes (El Morr, 2019b).

Kisielnicki (2017, p.5) highlights several problems that appear during BI implementation:

- a long development lifecycle and less visibility to user;
- users are not involved in the development cycles;
- after the design phase, there is no possibility to modify analytical requirements;
- testing is performed at the end of the development cycle, again without a possibility for change requests;

The computability approach is an epistemological criterion according to which, to be scientific, a theory of mind should be describable in the form of an algorithm. More precisely, the computability criterion is sufficient (even if not necessary) for a theory to be considered scientific. Two different meanings are defined for the computational approach:

- in a *weak sense* it is sufficient that it complies with the computability criterion;
- in a *strong sense* it is necessary that the problem be faced on the basis of the prescriptive epistemology of the three levels of Marr.

In its weak meaning, the computational approach has, in relation to psychology, a relationship somewhat comparable to that between mathematics and physics: the algorithmic description constitutes a formal language into which psychological functions can be translated.

Computational thinking is a set of transversal skills related to the foundations of computer science as a scientific discipline and means a mastering to the process of solving problems. The goal of computational thinking is to acquire interpretative perspectives of reality, which allows one to read the digital experience competently and responsibly.

Table 1. Main skills of computational thinking

Mental Processes	• *Algorithmic thinking* • *Logical thinking* • *Breakdown of problems* • *Abstraction* • *Pattern recognition* • *Generalization*
Methods	• *Automation* • *Collection, analysis, and representation of data* • *Parallelization* • *Simulation* • *Rating* • *Algorithmic problem solving* • *Storytelling*
Practices	• *Experimenting, iterating, tinkering* • *Testing and correcting errors (debugging)* • *Reuse and remix*
Transversal Skills	• *Create* • *Communicate and collaborate* • *Meta-cognition* • *Manage ambiguity* • *Reduce complexity*

As a comparison, we cite **Agile methodology**. The principles that guide the Agile methodology are presented in the Agile Manifesto (2001), a document in which are described four core values supported by 12 principles. The 4 core values are the heart of the Agile methodology:

(1) Individuals and interactions over processes and tools;
(2) Working software over comprehensive documentation;
(3) Customer collaboration over contract negotiation;
(4) Responding to change over following a plan.

Analysing the Agile principles, some perspectives between the development of projects using an Agile or Waterfall methodology (Kisielnicki, 2017) cab be highlighted. First, Agile is focused on people, Waterfall is focused on processes, on other words, one is "Customer Centric", the other is "Process Centric". This brings other relevant difference related to the requirements. In waterfall they are defined in the beginning of the project, in Agile they are based on business value with regular updates, which brings a performance measurement based on the value added to the business in Agile, against the conformity to the plan on Waterfall. Second, the initial planning is low in Agile and high in Waterfall, which brings a more adaptive, flexible, and responsive goal achieve in Agile when compared to a goal optimization focus on Waterfall. At last, in the problem-solving perspective, Waterfall is focused in selecting the best means to accomplish a given end through well-planned activities, and Agile is focused on learning through experimentation and introspection, constantly reframing the problems and the solutions.

Basically, four tools must be highlighted: User Stories; Epics and Initiatives; Product Backlog; Kanban Boards.

The ***User Stories*** are a method for representing requirements, using a template. As defined by Rehkopf (2020), the user stories serve several key benefits:

- Stories keep the focus on the user. A To Do list keeps the team focused on tasks that need checked off, but a collection of stories keeps the team focused on solving problems for real users.
- Stories enable collaboration. With the end goal defined, the team can work together to decide how best to serve the user and meet that goal.
- Stories drive creative solutions. Stories encourage the team to think critically and creatively about how to best solve for an end goal.
- Stories create momentum. With each passing story the development team enjoys a small challenge and a small win, driving momentum.

User stories are also the building blocks of larger agile frameworks like ***epics and initiatives***. In short, epics are large work items divided into a set of stories, and several epics can be aggregated in an initiative (Rehkopf, 2020).

The ***product backlog*** will be the list that will prioritize the user stories. The most relevant/urgent in the top, the less important in the bottom. This prioritization can have a defined rule that must be transparent for all. The product backlog will allow all the stakeholders to be aware of the list of "requests" waiting for development and their prioritization.

This product backlog can be the first column of a ***Kanban Board*** (Inflectra, 2020). The Kanban Boards are also one agile tool that must be considered, where a Scrum Board can also be an option. There are some differences between Scrum and Kanban boards, but they are superficial different, not relevant to the subject of this study. Despite the differences, the principles are largely the same. The essential idea is that with any of this boards the stakeholders will have transparency and a holistic visibility about the work to be done, the work that is ongoing, and the work already done.

Complex organizations, such as hospitals, generate and collect data to manage in large quantities. Laney (2001) observed that (big) data was growing in three different dimensions namely, *volume, velocity* and *variety* (known as the 3 Vs). From those, the "variety" dimension remarks on the different types of organized and unorganized data that any firm or system can collect. Furthermore, a lot of data is totally unused, and, in many cases, the extracted information is disseminated in an inefficient way. In fact, these data and information, if well used, are of great value both for the management and administration of the healthcare organizations, as well as for the treatment and experience journey of the patients, scientific research, among others. Therefore, continuous improvement strategies are urgently needed to gradually and effectively take advantage of the data generated, and to transform and distribute them in an equitable way. For this reason, BI systems developed in a sustained and evolutionary way are fundamental to make the best use of the full potential of big data in healthcare.

Implementing an agile project management methodology, for example, in a hospital context has significant challenges. The Agile methodology is different from traditional methodologies because Agile is focused on delivering small functional parts of the features as early as possible, constantly improving them and adding new functionality throughout the project lifecycle (Flora, 2014). Exploring the success and failure factors of an implementation of the Agile methodology is relevant to the success of a model.

The main objective is to investigate if an agile project management methodology, due to its characteristics, is best suited to the development of some components of a BI system in a hospital context in relation to other methodologies. It is specifically intended to investigate experiments carried out in these areas and to discover the Agile tools that can be used and impact the development processes of a BI system, considering the hospital context.

Agile methodology seems to bring tools that will allow to fulfil those needs with better collaboration between the developer team and the customers. Agile also brings more transparency and allowance to change. With the Kanban boards all the stakeholders can have the vision of the status of the project, perceive the product backlog priorities, and have better understanding of the impacts of changes in the overall project.

The incremental and interactive processes also bring a feeling of belonging to all the involved stakeholders. It is not just a product that is presented in the final of the project, possibly with amounts of changes to be done. With the continuous involvement of all the stakeholders during the development process, small improvements and changes can be done and the results can be used immediately.

HOW THIS BOOK WORKS

This book proposes transdisciplinary research, enabling progress towards an effective understanding of Problem Solving and Managerial Mindset Training, basing on Computational thinking methods, compared, in some case study, with Agile approach.

The analysis consists of key issues, problems, trends, and particularly new emerging ideas and innovation. Furthermore, it investigates questions, explanations of terms and concepts in order to probe further thinking.

This comprehensive and timely publication aims to be an essential reference source, building on the available studies on developing 'instruments of analysis' to train managers, analysts, politicians, and researchers.

The book consists of three sections and 13 chapters, which explore the adoption of computational thinking for individuals to gain managerial mindsets and successfully solve questions and problems in their domain of interest. This will include artificial intelligence applications, strategic thinking, management training, ethics, emergency managerial mindsets, and more.

This book is valuable for managers, professionals, practitioners, researchers, academicians, and students interested in how computational thinking can be applied for the training of managerial mind.

A brief description of each chapters follows:

Chapter 1: Artificial Intelligence, Human Intelligence, and Orientism Management (OM) Framework – A Way to Improve Managerial Mindset and Manage Emergency

The chapter intends to be a theoretical contribution for developers in the field of Artificial Intelligence, as well as a practical guideline for Leaders, as decision-makers, to manage tasks towards goal, and to understand what elements are useful to own personal and professional development to optimize performances. The proposed approach interprets the fluid nature of the decision-making process, looking at knowledge and knowledge activities as dynamic, adaptive, and self-regulative, based not only on well-known explicit curricular goals but also on unpredictable interactions and relationships between players. The knowledge process is emerging in human and biological, social, and cultural environments.

Chapter 2: Optimizing Supply Chains Through System Dynamics Modelling and Simulation – Lessons From the Navy

Through a case study, the Author describes how, in critical activities and organizations, decision making in face of complexity has been a growing normal. Complexity troubles humans due to cognitive limitations. Moreover, humans are merely able to understand cause-and-effect relationships that are close in time and space, not the paradigm of many complex socio-technical systems. Decision-making processes shall rely on models that help harness a problem´s associated complexity – among them the dynamics of supply chains. Models which typically fall into two broad categories - mental and formal models. Supply chains are complex systems, which may exhibit complex behaviour patterns. Decisions and policies within organizational systems are the causes of many problems, among them undesirable oscillations and other problematic patterns of the parameters of interest. A System is a grouping of parts that work together for a purpose. Hence the systems dynamics methodology is an adequate approach to deal with fuel supply chain management. A model was developed which helps manage Marine Gasoil supply chains in the context of the Navy.

Chapter 3: Risk Models and Managerial Strategies for the Global Supply Chain's Security – An Analysis of the Management of International Vulnerabilities and Emergencies

This chapter aims to investigate the effectiveness of managerial strategies to address sources of risk and strengthen global supply chains' security, with specific attention to disasters deriving from the interaction between extreme natural phenomena and pre-existing socio-economic vulnerabilities. The object of study is represented by the mutual learning opportunities that can arise from socially-oriented cross-sector collaborations between humanitarian agencies and enterprises. With the ultimate scope of grasping the economic impacts, the business perspective provides several insights to integrate a proactive approach into the corporate strategic planning to ensure a capacity to react with respect to external stimuli. In particular, an in-depth study was carried out dedicated to the regulatory standards that allow the development of this management system. Finally, private sector's involvement in humanitarian programs is analyzed in order to highlight the benefits in terms of effectiveness of operational performances but also of safety of communities' critical infrastructure.

Chapter 4: State of the Art of XR Management Training Applications

The Author focuses on the use of high technology to train managers. Recent advancements in Extended Reality (XR) immersive technologies provide new tools for the development of novel and promising applications for business. Specifically, Extended Reality training applications are becoming popular in business due to their advantages of low cost, risk-free, data-oriented training. Extended Reality Training is the digital simulation of lifelike scenarios for training purposes using technologies such as Virtual Reality, Augmented Reality, Mixed Reality. Many applications are already available to train employees to develop specific technical skills, from mainteinance to construction. The purpose of this chapter is to review the emerging XR applications developed for management training. Specifically, this paper will focus on the training of some key skills in management such as leadership, problem solving, emotional intelligence, communication, and team working.

Chapter 5: Professional Mindset in Troubled Times – Creative and Collaborative Sharing in Epistemic Community

Globally, COVID-19 have stressed social and personal tensions in professional life. This paper focuses on the networked dimension to highlight the workers need for social connections. Some suggestions are proposed to implement an epistemic community to sustain creative and collaborative professional development in disruptive time. Two points are discussed to scaffold epistemic communities in the organisation: a) the inter-professional dimension to embrace complex topic; b) the emotional dimension as resources to embrace professional transformation.

Chapter 6: Learning From the Past – Intelligence in Decision-Making Processes During Prisoners of War and Hostage Crises

Starting point of the Author is that learning from past experiences should be considered vital in every decision-making process. It becomes even more important when lessons could save lives. Past experiences could indeed support the development of survival courses and reintegration processes for military and civilian personnel deployed in hostile environments. Moreover, the integration of lessons from the past and basic techniques could be useful for anyone working, or simply traveling, in areas where the risk of isolation and/or kidnapping is considered very high. During an isolation event, every aspect should be considered, from support to families during captivity to the reintegration into everyday life. Governments, NGOs, and international organizations should look at the past and learn how to establish an effective system to repatriate and reintegrate their representatives.

Chapter 7: Symbioses as an Alternative to Master/Slave for Artificial Intelligence

A very interesting scientific interpretation is described by this Author. Human society is facing enormous problems this century as result of Our Climate Crisis. These problems include sea level rise and the loss of farming capability. Society will need all the new tools it can develop to address these problems. Artificial Intelligence with Deep Learning is one of these powerful tools, and it is new. Exactly how it will be used has not been determined. The current approach to the human/AI interface is referred to as Master/Slave. The human simply tells the AI what to do. This arrangement has many problems, and replacing it has been suggested. One possible new arrangement is a Human/AI symbiosis. This would require a long-term relationship between a specific human and a specific AI. A novel, Born to Storms, exploring this arrangement is discussed at length.

Chapter 8: Artificial Intelligence, the Risk to Do Something Against the Man – Innovation Must Be Governed to Avoid Disgregation

The Author proposes a very critical point of view of the issue. A long time ago we started to speak about Artificial Intelligence. It was like a dream for many years and now, with the tremendous computing capability and data availability, we are living in a kind of reality where we think everything is possible because we can train machine to do more or less everything man is able to do both with his hands and with his brain. But we must keep in mind that we can train machines to do what we do, in the way we do, but it is quite difficult, for example, to train machines, and not only, to take decision in the way and with the quickness usually necessary. And what about ethics? What kind of future can we design?

Chapter 9: Knowledge Obsolescence and the Future of Work – Relevance of Knowledge and Impact to Jobs

The chapter delves into a range of influencing factors that are governing individual and corporate behavior, driven both by changing human circumstances - economic, social and environmental - as well as rapid changes to organizational cultures and endeavors - models, markets, governance, customers, competition. A veritable confluence of these factors is impinging into workplaces in a never-before seen manner, particularly in its inherent complexity, and constant change. Jobs are being redefined, created and eliminated at the same time, putting significant pressure on individuals seeking to pursue careers. Alongside, knowledge acquired over a certain period is becoming quickly obsolete, resulting in people having to shift gears quickly. Most fail, with consequences impacting both social structures and organizational cultures. Individual behavior is significantly deteriorating toward colloquialism driven by a sense of victimization. How do we address all these challenges and stay on top of the future? This chapter's aim is to distill into the answers to this question.

Chapter 10: Computational Thinking and Didactics – New Horizons

The Author proposes a view of the didactic perspective. The failures and problems of school education today are caused by many factors, stemming from the bias of behaviorism that the teacher is burdened with, as well as being the result of changes in the way of thinking of new generations, often caused by the negative impact of technology - its source is usually in the process of teaching and growth itself. Computational Thinking allows us to take complex problems, understand what the problem is, and develop solutions. We can present these solutions in a way that both computers and people can understand.

Chapter 11: Models for Strategic Thinking in Intelligence Analysis

The Author explores the relationship between strategic intelligence and computational approach. Intelligence as a field of activity that has a rich history, yet it still lacks clarity of definition and agreement. This is not to denigrate the performance of many of its practitioners; rather, these comments are designed to point up the fact that, despite the length and breadth of its historical practice, there is still much to do to explore the boundaries, opportunities, and limitations of the application of intelligence in the world of enforcement. The Cultural Roots of Strategic Intelligence is compelling and urgent not just for those involved with the professional intelligence community, but also for anyone seriously committed to interdisciplinary studies, cross-cultural understanding, and most importantly, to the development of a rigorous discipline of politics as well as cultural genetics. But a strategic planning does not guarantee strategic thinking.

Chapter 12: Unmanned Aerial Vehicles – Their Use During the American Presidencies of Bush Jr. and Obama

The Author proposes a specific case study, based on the use of Unmanued Aerial Vehicles to manage critical political situations. Last January, the General Secretariat of the Council of the European Union produced an estimate that, by the year 2035, the drones' sector will create up to 100,000 new jobs and, in addition, will have an economic impact of more than ten billion euros per year. This chapter takes its cue precisely from the central importance of Unmanned aerial vehicles, commonly known as drones, and tries to sum up their future. Indeed, despite the enormous success of drones over the decades, many are the criticisms related to unmanned aircraft on board. The first paragraph aims at outlining an historical framework, while the second paragraph analyses the impressive role played by drones within the counterterrorism policy of the US presidency, with particular regard to Barack Obama's mandate. The third paragraph applies a SWOT type analysis to UAVs, taking into account both their advantages and disadvantages. Finally, the fourth paragraph investigates the Human Factor problem within UAVs, which was inspired by Dr. Giovanni Miranda's PhD thesis.

Chapter 13: Ethics and Managerial Mindset in Politics

The chapter is focused on the strategy model applied in politics. Though the specifics of a strategy model vary greatly depending upon the nature of the model, the basic definition of such a model remains the same in all industries and sectors, such as in policy management. Basically, a strategy model constitutes a strategic plan, or model, designed to improve a process, and it need a specific mindset by developer. Organizations use strategy models to improve operations and meet their goals. The development of such a model requires identifying the key goals of an organization, identifying the key components or steps. of the process leading from a starting point to the achievement of those goals and creating ways of maximizing the process for optimal output and the achievement of the stated goals. Strategy models are of varying degrees of complexity.

REFERENCES

Austin, N., & Rankov, N. (1995). *Exploratio: Military and Political Intelligence in the Roman World from the Second Punic War to the Battle of Adrianople*. Routledge.

Cao, X. J., & Yang, X. F. (2013). Application of Business Intelligence in Hospital Information Management. *Applied Mechanics and Materials*, *373–375*, 1098–1101. . doi:10.4028/www.scientific.net/AMM.373-375.1098

El Morr, C., & Ali-Hassan, H. (2019a). Healthcare, Data Analytics, and Business Intelligence. In *Analytics in Healthcare. SpringerBriefs in Health Care Management and Economics*. Springer., doi:10.1007/978-3-030-04506-7_1

El Morr, C., & Ali-Hassan, H. (2019b). *Analytics Building Blocks: A Practical Introduction*. doi:10.1007/978-3-030-04506-7_2

Evelson, B. (2008). The Forrester Wave™. *Enterprise Business Intelligence Platforms*, *Q3*, 2008.

Flora, H. (2014). A Systematic Study on Agile Software Development Methodlogies and Practices. *International Journal of Computer Science and Information Technologies*, *5*, 3626–3637.

Inflectra. (2020). *What is Agile Kanban Methodology?* Retrieved from https://www.inflectra.com/methodologies/kanban.aspx

Kisielnicki, J., & Misiak, A. (2017). Effectiveness of agile compared to waterfall implementation methods in it projects: Analysis based on business intelligence projects. *Foundations of Management.*, *9*(1), 273–286. Advance online publication. doi:10.1515/fman-2017-0021

Laney, D. (2001). *3D data management: controlling data volume, velocity, and variety, Application delivery strategies*. META Group Inc.

Manifesto for Agile Software Development. (2001). Retrieved from http://agilemanifesto.org/

Mettler, T., & Vimarlund, V. (2009). Understanding business intelligence in the context of healthcare. *Health Informatics Journal*, *15*(3), 254–264. doi:10.1177/1460458209337446 PMID:19713399

Monod, J. (1970). Le hasard et la nécessité. In *Il caso e la necessità Saggio sulla filosofia naturale della biologia contemporanea*. Edizione Mondadori.

Monteiro, M., & Gonçalves, B. (2021). *Develop Business Intelligence systems in hospitals: Study for an Agile Project Management approach* (Master's dissertation). Polytechnic of Porto.

Rehkopf, M. (2020). *User Stories with Examples and Template*. Atlassian. Retrieved from https://www.atlassian.com/agile/project-management/user-stories

Chapter 1
Artificial Intelligence, Human Intelligence, and Orientism Management Framework:
A Way to Improve Managerial Mindset and Manage Emergency

Luisa Dall'Acqua
University of Bologna, Italy & LS TCO, Italy

ABSTRACT

The chapter intends to be a theoretical contribution for developers in the field of artificial intelligence. It also means a practical guideline for leaders, as decision-makers, to manage tasks and optimize performance. The proposed approach interprets the fluid nature of the decision-making process looking at knowledge and knowledge activities as dynamic, adaptive, and self-regulative, based not only on well-known explicit curricular goals but also on unpredictable interactions and relationships between players. The knowledge process is emerging in human and biological, social, and cultural environments.

INTRODUCTION

How Artificial Intelligence technology is processing has its roots in the mechanism by which our cognitive process works. The core is "what do we mean with Intelligence?" The two classical definitions of Intelligence are:

DOI: 10.4018/978-1-7998-7126-2.ch001

- the ability to solve new problems and adapt to the environment
- the ability to design and implement complex purposes

In reality, everything we do, no matter how simple it seems to us, is the result of the extraordinary assembly of billions of brain cells.

It seems natural to us to go back to our childhood memories, as well as to live the fantasies of our dreams or be aware of colors and sounds existing in the world.

Imagine carrying billions of stars inside our head, calling them neurons and connecting them together you will have the exact representation of our brain.

Figure 1. Neurons

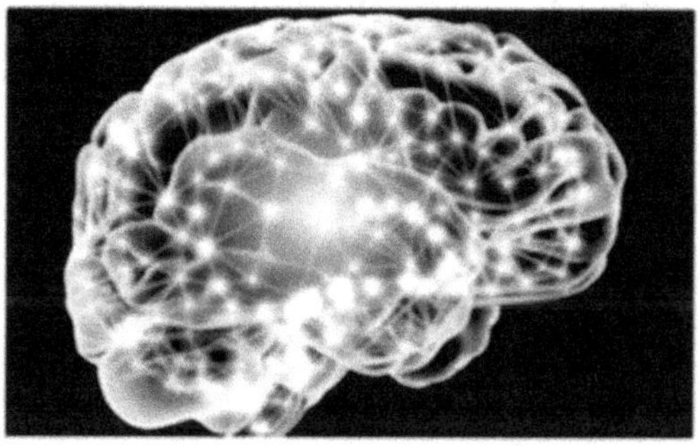

From the action of all those stars, everything (that characterizes us as human beings) is born, who we are, how we act, why we fall in love, our consciousness, our thoughts, basically what we call mind.

The hard problem, is right here, in this leap from atoms to thought, from neurotransmitters to consciousness. This is what allowed man to think about the precision of the laws of physics, to produce the emotions of Bach music or the extraordinary beauty of a Raphael painting, or to rise to the idea of the existence of God.

Today there is a lot of speculation about a possible future in which humanity will be overtaken or even destroyed by machines. There are cars that drive themselves, of big data, and even ***transhumanism,*** the idea that it will be possible download our experience and awareness to a computer and live forever.

Wetware

Wetware is a term drawn from the computer-related idea of hardware or software, but applied to biological life forms. Especially the central nervous system (CNS) and the human mind (the previous words was Meatware, coined in the nineteen-seventy years).

Who think it will be possible to achieve machines that will surpass humans, believes that consciousness only emerges from the brain, that it is produced by something like the software that runs in our computers.

Therefore, with more sophisticated software, our robots will end up being aware. But is this really possible? For example, it is certainly possible to build a robot capable of recognize the particular combination of different molecules emitted from a rose that allow it to correctly identify the rose by its smell. That robot wouldn't feel any sensation. It would not be aware of the smell in the form of sensation. To be aware you have to smell the scent. The day when artificial intelligence will be able to approach human intelligence is near, but its abilities must be judged on three levels of cognitive ability:

- to see (association)
- to do (intervention)
- to imagine (counterfactuals) as a plastic change

Currently, artificial intelligence is based on the first level. The "doing" is simply a special way of an association between a situation and behaviour. The robot stops at electrical signals, and from those signals it can generate what corresponds to the name "rose", just as a symbol, which can also be associated with some kind of action.

The brain does much more, because we smell the rose, and through that feeling we connect in a special way to that rose and the meaning that roses have in our life.

For example, the factors that led to the idea of ai could be better than a man, were essentially two:

- the first one is the incredible forecast of growth in the computing capacity of computer systems, for which, according to Moore's law, *in the year two thousand and forty-five a single computer could have the computing power of the whole of humanity*
- the second factor arises from knowledge about the human brain, as the result of an evolution lasting millions of years. at birth, its structure is already decided by the genes, by the dna we have inherited.

We have our ***hardware***, for example the set of information in our DNA that constitutes the basic structure of the brain and provide us with the methods of use, almost ready for birth, fixed, practically stable over the last fifty thousand years, and with a low data storage capacity (about 1 Gb).

On the contrary, our ***software***, for example the set of ductile and modifiable neural networks which will allow the development of a large part of intelligence, is developed at a later time, through a learning process that, throughout life, continuously transforms our brain.

This is what makes us have a much higher intelligence than that transmitted by DNA alone, and an information storage capacity of up to 100 terabytes.

The challenge that scientists faced with ai was to be able to build a brain, that, in addition to having ductile software, also had hardware that could be implemented at will, so as to have an immensely powerful machine capable in a few decades of an evolution that (for biology) brain would take thousands of years.

ARTIFICIAL INTELLIGENCE'S STRENGTH POINTS

A.I. beats us in everything related to speed of operation and number of computational elements.

Artificial Intelligence is able, with much, much more efficiency than man, to detect significant structures within even very large databases. The A.I. therefore had to learn to solve problems which cannot be described mathematically that the brain easily solves in an intuitive (automatic) way, such as recognizing words or faces in images.

Since there are no rules for this activity, it was necessary to learn from experience. If knowledge comes from experience, computer can do without human intervention. In fact computers can function in terms of hierarchies of concepts, where each concept can be defined with its relations to simpler concepts, and the computer can learn new complex concepts by building them from the aggregation of simpler ones that it already has.

A.I. Initially set out to solve practical problems, that is

- automated production process routines
- recognize text or images
- make medical diagnoses
- support basic scientific research
- other similar tasks

Figure 2. A comparison between the human brain and supercomputers

HUMAN BRAIN	SUPERCOMPUTER
90 billion neurons this number cannot be changed	large as a hangar or as several hangars many high-speed cables as you want
38 billion operations per second	the HPC5 70 million billion operations per second
a brain today functions roughly like a brain from 1944	An iPhone X outperforms a 1944 IBM computer by 4 million billion times in computing power
much of the structure of our brain undergoes physiological deterioration and dies	a computer can be continuously deployed and perform an indefinite number of downloads

Over time it became clear that the real challenge is that the A.I. had to face was to acquire the own knowledge, by extracting structures ("patterns") from raw data external.

This ability is called "machine learning".

Machine Learning

Machine Learning can be defined as software that creates algorithms (a sequence of actions) that improve its performance in course of time in the context of a specific task, after having carried out a series of experiences.

There are at least three macro-areas of learning methods, which differ in how they treat the attribute to be modeled, namely the target class or target score:

Figure 3. Macro areas (source: https://www.developersmaggioli.it/blog/machine-learning-la-scienza-delle-decisioni-automatiche/)

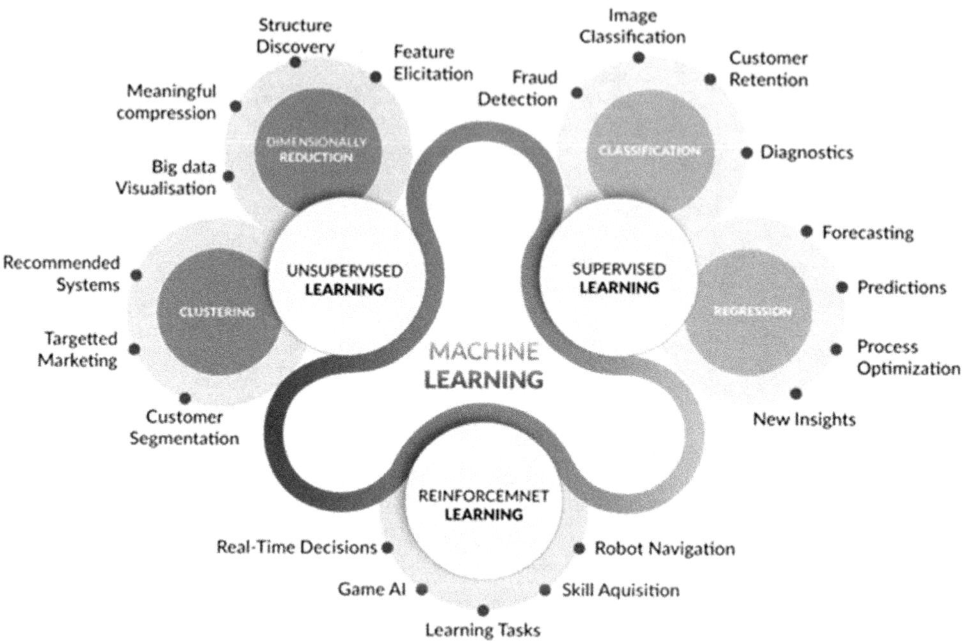

Figure 4. Description of the 3 marcoareas

TYPE	USE	APPLICATION
supervised learning	uses a series of attributes to approximate target score or target class in regression or classification models	typically gets very good models for **predictive analytics**
unsupervised learning	includes processes to extract associative regularities between attributes in the form of rules	very useful for **data science**
reinforcement learning	target class or target score are generated by a model that updates itself by interacting with new data	in the case of **human-machine interaction** or **gaming**, or **robotics**

Deep Learning

Recently, GPUs based on deep learning technology have set the stage for the new era of computing in which the GPU acts as the brain of computers, robots and self-driving vehicles capable of perceiving and understanding the world.

Deep learning is a supervised process that automates the extraction of large amounts of attributes from data and then classifies them in a supervised manner by means of neural networks. It is applicable in many types of situations, for example in computer vision and speech recognition, or in industrial applications, but much less in data science

- it is an automatic learning that requires long causal chains of computational phases.
- it is a process that continuously evaluates and assigns credit to many interconnected computational phases (such as the human brain) towards the desired outcome

A question: does the crowing of the cock cause sunrise? the answer is obvious if you are a human being, but a machine can only understand that the crowing of the cock and the sunrise are related, not that one causes (or not) the other.

Causal inference or causal reasoning is the most representative inference of human intelligence because it implies the prediction of the effects of one's actions. Causal explanations and conditions, not mere facts, make up the bulk of our knowledge.

Figure 5. Subsets

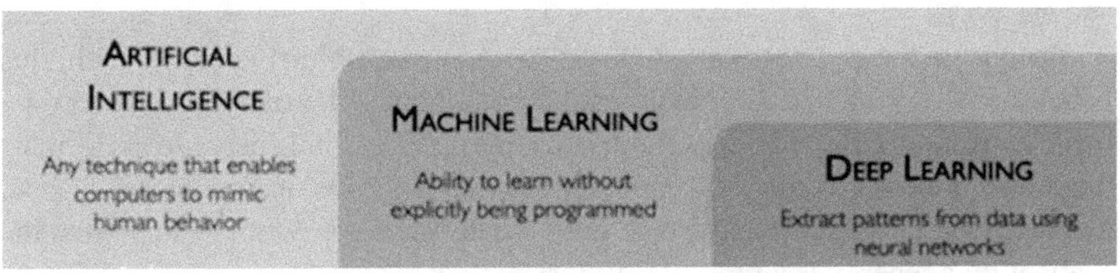

How do human and artificial intelligence deeply differ? "in the ability to ask questions and find answers". But, contrary to what one might think the most important ability is to "ask questions", the problem finding (posing and setting).

BRAIN'S (STRENGTH POINTS)

Plasticity

One of the most important and peculiar characteristics of the human brain is plasticity, which consists in the ability to continue to change and learn throughout the course of life, up to senescence.

Every time we listen to something that interests us, or after reading a few pages of a book, something moves between our neuronal circuits, the brain expands, synapses change, knowledge is formed. It is a very personal and necessary imprint that is affected by the experiences and environment in which we live.

In other words, it will be the culture that each of us will accumulate that will contribute to developing the brain of every single person, thus making each of us a unique entity, different from any other, developing and improving, year by after year, throughout the life. life, mental faculties that will allow us to reach the highest possible degree of intelligence.

Empowerment

The brain, although so complex, in its functioning follows simple principles, which are those that evolution has imposed on it, in order to function well, and survive.

From birth we follow two rules (more or less opposite).

On the one hand, ***a push*** that leads us towards exploration and which determines the accumulation of knowledge. But on the other hand, we have a brain software, programmed to constantly ***pull*** the brake on our cognitive activity and save work and energy.

Energy saving is a necessity, that the brain must absolutely comply with. this is because every brain action requires a consumption a lot of energy. The brain, while constituting 2% of body weight, consumes 20% of total energy to perform its functions. If the set of synapses carries out about 38 billion operations per second, and the brain never stops and works even when we sleep, we understand that the availability of energy is almost always just enough.

So, a strategy.....

Unconscious

Even if the number of neurons seems enormous, in reality it is insufficient for all the necessary mental operations. An extraordinary example of energy saving is the unconscious. It is estimated that about 95% of daily brain activities take place in the unconscious. Far from a psychological point of view, the true is that in about 100 million neurons, of the million of billion connections, our consciousness controls only 5%; most of the work is occult. This mechanism frees the mind from a big amount of processing and information that remains in the subconscious.

Wonderment

Another peculiar characteristic of the brain is that it is bored, and loves to be continually stimulated. In front of a news, brain activity involves a large number of areas, and connects groups of neurons together according to new patterns. In practice, the brain (to be activated) wants to be amazed, it has to get excited.

Emotions

Rationality has its seat in the prefrontal cerebral cortex, which appeared about a hundred thousand years ago, while emotions are based in a set of much older nuclei, those of the limbic system, which developed two hundred-three hundred million years ago. What's their role?

Emotions are not just feelings, they are (mostly) survival algorithms, evolved

- to let us avoid unpleasant situations (threats, dangers, events that scare us)
- to favor pleasant emotions (people or situations that make us feel safe and that cause us pleasure).
- to give strength to our memory.

We memorize everything that excites us. The brain is very selective, it does not waste its limited neural networks and limited energy to fix useless things in the mind. Emotions also play an important role in every decision-making process, in the choice of good and evil. And a computer lacks this prerogative.

Thought

Another aspect of human intelligence is thought. Even in the organization of this fundamental function, the human brain oscillates between emotions and rationality. In what way?

We can consider 2 systems, and both of them are complementary ways of operating.

System 1, the fast and intuitive thinking, is activated in a particularly serious and dangerous situation, where having a quick response is essential and because the mind wants to avoid complicated analyzes, and uses mental shortcuts, intuitively, even if it can lead us to cognitive distortions. A big issue is the possibility of cognitive biases.

So, the mind activates ***System 2***, the slow rational thinking. It involves the cortex of reasoning and requires conscious reasoning. The mind activates slow thinking in making its decisions, it wants to sift through all available information, and use the nervous system in its most developed cognitive properties.

Figure 6. Brain's systems

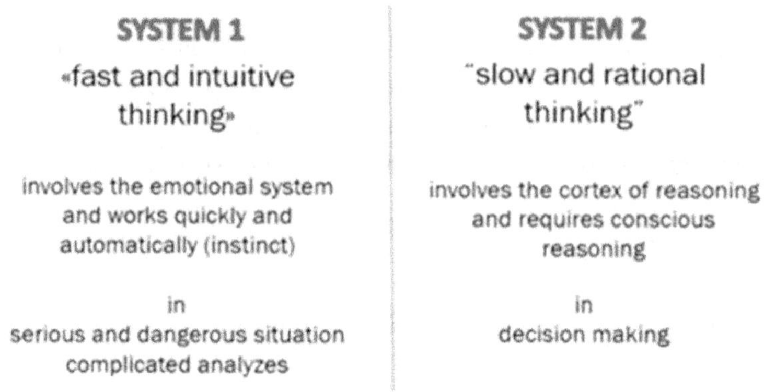

From an energy point of view it is a very expensive way of operating. for this reason the brain, which must always save energy, tries to privilege fast thinking (the instinct) and follow mental shortcuts, basing on hidden elaboration of experiences.

Creativity

According to some, computers, once they become powerful enough, will be able to display the same kind of creative intelligence attributed to people. But if we imagine creativity as the expression of a free association of ideas, thoughts and emotions, a computer (driven by rigid algorithms) will never be able to express it. Furthermore, only we have what Marcus du Sautoy, a professor of mathematics at the university of oxford, defines: the "***human code***"; that is, that extraordinary ability to imagine, renew and create works of art that elevate, expand and transform what it means to be human.

Consciousness

Consciousness is such a complex feature. It implies so many things, such as awareness of oneself and of the reality that surrounds us. Just to explain: the event of a glass that crashes on the floor has surely a cause and an effect, but also a "condition": the fragility of the object. Thinking of us strong, weak, is an awareness action. To understand this concept is too hard for a computer to date. As well as a moral vision of the world, thinking and the ability to reflect on one's thoughts, develop a project on these, and so many others. It is difficult that a raw material, only by having reached a level of high computational complexity, can make it emerge.

THE RESEARCH, TOWARDS A NEW KNOWLEDGE BASE'S IDEA

The present work describes a possible view of research.

Figure 7. Three models

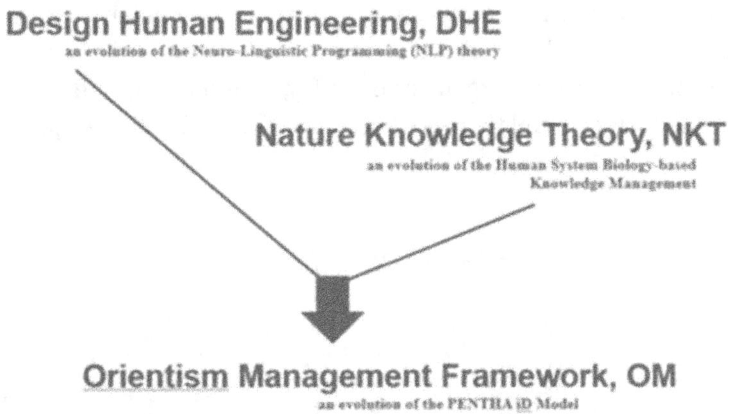

Design Human Engineering, DHE (by R. Bandler) offers the opportunity to design new internal states and new strategies, which are dynamic and engaging at the same time. Just think about how many positive states you can experience. however many there may be are still not enough, you can always do better.

The underlying premise of DHE is that whatever an individual has available now is still a long way from anything he is potentially capable of achieving (Bandler, 1996; Bandler et Al. 2013).

Figure 8. A comparison between NLP e DHE (by R. Bandler)

Neuro-Linguistic Programming	Design Human Engineering
NLP is based on the replication of human behavior (modeling)	DHE is based on the creation of new models
NLP studies the sequence of access to certain information	DHE takes into account the simultaneity of how we process information
NLP is concerned with dividing human behavior into smaller and smaller pieces	DHE is the model through which larger and larger pieces of human experience are put together to achieve success

Nature Knnowledge Theory, NKT (by Md Santo) is focused on designing a semantic interpretation of the knowledge process. It is a transdisciplinary approach (Md Santo et al., 2014), through contextual interaction of km components. The organization is interpreted like a living mechanism, having interior own consciousness, wherein each agent works with own specific profile, having to be orchestrated in a "nature knowledge continuum".

The main goals are: an establish taxonomy of enterprise process business (indicating complexity aspect of risk management) based on process classification framework (PCF method and proceed to strongly recommended community of practice (COP) and learning organization (LO), as modern trending issues before developing it to further usual themes of management tools (dall'Acqua&Md Santo, 2012, 2014).

Figure 9. Nature Knnowledge Theory, NKT (by Md Santo)

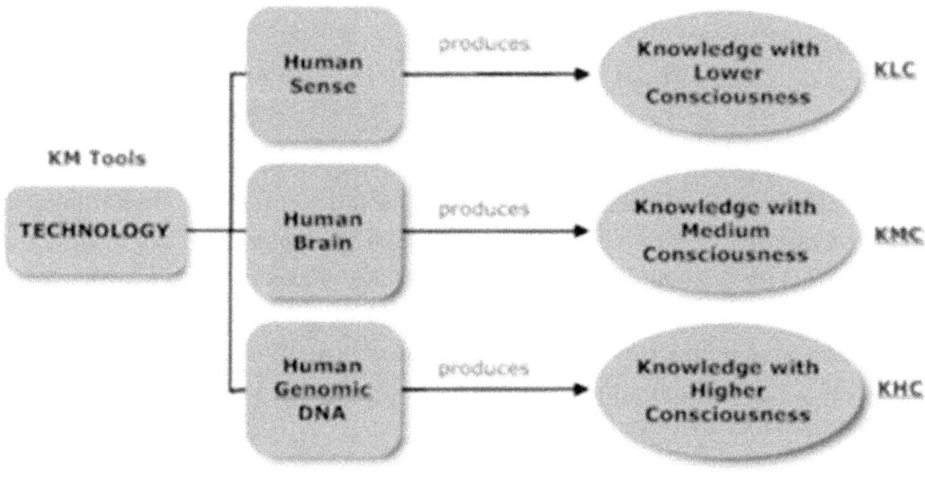

Orientism Management (OM) view (by L. dall'Acqua) is a new conception of the knowledge management: it moves towards a knowledge awareness, extension of own horizon points of perspective and meanings management, in a network of relationships, according to the complexity theory and connectionism, starting from own strength points and profile, and consciousness transfer in a knowledge interface phenomenon.

New variables, factors and criteria are designed to direct and motivate the mind in the decisional process, in self-reference, to be considered in a lifelong guidance to have success in choices, assumptions of responsibilities and achievement of objectives, over the uncertainty (dall'Acqua, 2014).

Figure 10. A mash between Orientism Management and PENTHA model (by L. dall'Acqua)

Following, this last model is described to allow the design of a managerial mindset training, as well a possible way a new concept of knowledge base framework.

ORIENTISM MANAGEMET FRAMEWORK

The current technological advancement has multiplied the difficulty of managing an innovative business system, in changing to guarantee training of managers, entrepreneurs and researchers to become real decision manager in a complex multi-referenced digital world. . It is often said that to digitize a company it is necessary to: introduce new technologies, implement current ones, modify business models and processes. Many quantitative and data-centric problems can be solved using computational thinking, understanding it provides a foundation for solving problems that have a social impact in the real world. Do we want to have business processes digitized or to be digital? They are two profoundly different concepts. So in the business environment, to excel it is necessary not only to invest in digital skills but also (and above all, I would add) in leadership skills..

This analysis refers to the Orientism Management concept [dall'Acqua, 2018] to describe Leader's risk management.

Key points of Orientism concept are:

- The creation of new models of behavior
- The simultaneity of how we process information
- Larger pieces of human experience are to be put together to achieve success
- The complex interconnection between reference points

New variables, factors and criteria are needed to direct and motivate the mind in the decisional process, in self-reference, to be considered in a lifelong guidance to have success in choices, assumptions of responsibilities and achievement of objectives, over the uncertainty.

OM is composed by intertwined 10 Knowledge Managements (KM) typologies, each based on areas of development and improvement of own personal leadership: KM with consciousness (3 levels), meaning, feeling, will, understanding, personalization, availability, synergy.

Figure 11. KMs of OM Strategy (by L. dall'Acqua)

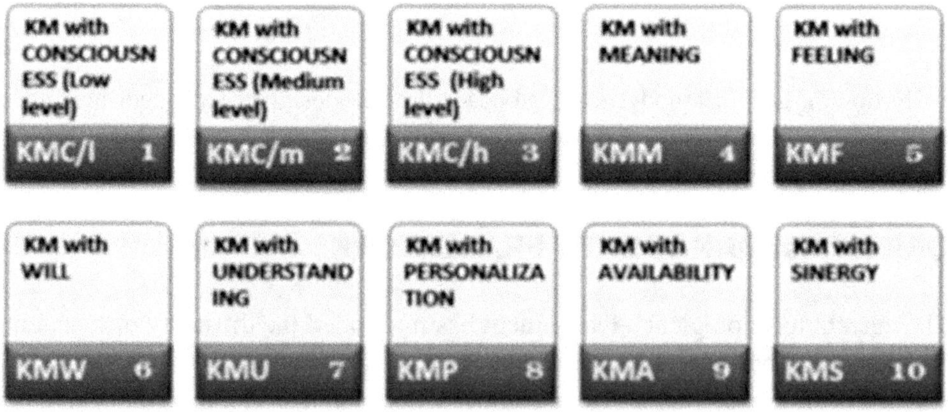

The outcome is the design of a training strategy for Leaders.

1ST KNOWLEDGE MANAGEMENT WITH CONSCIOUSNESS (LOW LEVEL) - KMC/l

As Orientism component, it focuses on the learning "***TO KNOW WHAT***" by the Leader. It means to activate a *prescriptive plan*, that concerns a knowledge base of updated regulatory guide, according to own organization profile and needs.

SWOT Analysis

- ***Strenght points***

Ssolid identity and high professional identification

- Compulsory education
  - ***Weakness points***
- Increansgly complex organizationss
- Deep change in the management paradigm
  - ***Opportunities***
- Leader as crucial agent of the innovative educational process
- Getting better remuneration ?
  - ***Threads***
- Excessive (or contradictory) legislation
- Low financial resources to invest (needs of fundings)

Training to Improve KMC/low

In PENTHA model approach, training for Leaders is to apply ***FORMING*** mode**:** a method concerning a) a topic introduction to make pre-service Leaders aware of the organizations reality, introducing to the role through normative, with a robust case of studies to apply it, and b) an increasingly updating of law for Principal in service. The training purpose is to empower attitudes, according to vision, professionality and dedication to the role.

2ND KNOWLEDGE MANAGEMENT WITH CONSCIOUSNESS (MEDIUM LEVEL) - KMC/M

As Orientism component, it focuses on the learning "***TO KNOW HOW***" by the Leader. It means to activate a *management plan*, concerning how to provide efficacy and efficiency by a distributed management.

SWOT Analysis

- *Strenght points*
  - needs of leadership paradigm change
- *Weakness points*
  - multiplicity of responsabilities
  - mistakes of leadership
  - Employees coordination in relation to the process of change
- *Opportunities*
  - Organizations middle management
  - Organizations digital promoter
- *Threads*
  - low or not adeguate human resources

Distributed Leadership: Middle Management

Speaking about top management includes the need of circulation of information top down and bottom up, and connection by all members, focalization on quality of contributions, and valorization of critical, creative and strategic thinking.

The “coadjutors” are not simply “collaborators” of the principal: they support the executive management of the organizations in various ways and are the real “middle management”.

To apply this organization principals have to manage a strategic use of delegation (supposing sufficient incomes for their additional workload, besides managing).

This means several main possible criticalities of management, such as: absence of candidates, not competent candidates, not competent but available candidates, competent but insecure or embarrassed candidates from the context, leadership mistakes.

To analyze the problem, this study refers to a few historical leadership theories, evaluated among the most suitable theories: Situational Leadership of Hersey and Blanchard [1982] and Formative / Supporting Leadership of Black and Moudon [1994].

Case 1: Unqualified Person

It is a common situation in a organizations, and often principal has not choice to delegate duties.

If coadjutors are unqualified persons for the required role, so they need precise instructions on the job to be performed, and specific knowledge of legislation and procedures.

The principal takes on the directive role of "*telling leader*", who makes decisions for the coadjutors, plans and organizes their work.

Criticism in criticism: keeping active the relational level, not to risk overcoming too much the coadjutor's initiative.

Case 2: Available Person, But Not Yet Competent

It's the case of a person, willing to professional development, to take responsibility, but not yet able to do the assigned work.

The principal takes on the role of "*coaching leader*", who has the responsibility of the decisions, but explains and motivates them, with the aim of actively involving the collaborator

It makes the relational level high.

Case 3: Competent but Insecure Person

It's the case of a competent coadjutor, but uncertain for fear of liability, non-inclusion in the context, or no sense of belonging

The principal takes on the role of "*empowering leader*", who is participative, decides together with the collaborator the decisions to be taken to facilitate the work of the coadjutor.

The relational level is high, in support, reassurance, empowerment, trust.

Case 4: Competent, Responsible, Reliable, Self-Confident Person

It's the case of an autonomous coadjutor, able to set goals, following the principal's guidelines.

The principal takes on the role of "*delegating leader*", who defines the general reference work-guidelines, communicates willingness to be of help and support, supervises and controls the results (not the process).

It keeps the level of interaction within limits.

Leadership's Possible Mistakes

To be a coach leader means to direct managerial action not to the achievement of the assigned work result, but to the development of the skills and attitudes of the team members. It means not just speaking about goals, but also and above all how to achieve these goals. It means replacing support, team management, and personalized development, with push and pull logic.

Managing the previously described main situations, the principal can fall in the following common behavioral mistakes:

- ***Connected with the level of maturity of the staff***
    - To assume a style too *directive*, risking to become *authoritarian*
    - To assume a style too *supporting,* risking to become *paternalistic*
    - To assume a style too *persuasive*, risking to become *handling*
    - To assume a style too *participative*, risking *paralysis* in the relationship
    - To assume a style too *delegating,* risking to assume an attitude of *"passing the buck"*
- ***Connected with the public image or perception of the principal***
    - Perceiving a *self-centric leader*, rather than coadjutor's centric relationship and their organizational growth; or a *failed leader* (laxism).
    - On the other hand, *not* perceiving *level evolution signals* in coadjutors, or signs of inadequacy of an assigned role
- ***Connected with the judgment of people***
    - Employees, always classified into two categories: motivated and reliable or lazy, irresponsible, incapable; competent (left in peace) or incompetent (always checked)
    - Employees, always classified equal, at medium level, causing not sufficiently understanding or minor management of crucial situations (between motivational support and professional reinforcement)
- ***Connected with the principal attitude***
    - With tension "to run the game", for needs to feel "indispensable", and, usually with a very specialized vision of the organizations and no graduation in the assignment of tasks
    - Feeling the need to get rid of the coaching role, with the consequent risks of overloading the available collaborative employee, or too early delegating on, and leaving alone the coadjutor, without guidance, resulting in loss of control
    - With high expectations of results, leaving crucial situations in the hands of people who are not adequately competent but available, and the possible consequent disappointment or damages without forecasting

Training to Improve KMC/Medium

In PENTHA model approach, training for pre-service Principals is to apply ***MODELING, PERFORMING*** and ***SCAFFOLDING*** modes. Modeling is a method for which the Tutor demonstrates how to perform a task It means give a robust case of studies to the Leader. Performing is a method to push Leader on motivation and attitude in a performance. Scaffolding is a method with focus on favoring the adaptation of the knowledge/learning path taken, a reflection on the actions developed by the Leader, stimulated by the Tutor

The training purpose is to empower skills, effective awareness of the role and work outcome, according to professionality, vocation and mission of the role.

3RD KNOWLEDGE MANAGEMENT WITH CONSCIOUSNESS (HIGH LEVEL) - KMC/H

As Orientism component, it focuses on the learning "***TO KNOW WHERE***" by the Leader. It means to activate a *descriptive plan*, concerning contextualized design and action

SWOT Analysis

- ***Strenght points***
  - to can control the management process as "action pin"
  - Knowledge of the territory and its resources also for guidance activities
- ***Weakness points***
  - Work on multiple sides
- ***Opportunities***
  - Private fundings
  - top down a bottom up support
- ***Threads***
  - to identify critical contextualized relationship

Training to Improve KMC/High

In PENTHA model approach, training for Leaders is to apply ***NARRATING*** mode. It is a method concerning two different aspects: a) introduction of the topic to attract the attention of Leader, appreciating of different leadership styles and different forms of intelligence; b) to encourage Leaders to verbalize their experiences, not only for dutiful reports, but also in a personal e-portfolio. The training purpose is to empower skills, effective actions and work outcome, loving the profession and according to professionality, mission and passion of the role.

4TH KNOWLEDGE MANAGEMENT WITH MEANING - KMM

As Orientism component, it focuses on the learning "***TO KNOW WHY***", that the Leader needs to manage. It means to activate a *social plan*, concerning a reasonable motivation for decisional actions in terms of efficacy and efficiency.

SWOT Analysis

- ***Strenght points***
  - solid regulatory system
- ***Weakness points***
  - Lack trust and conviction in the Reform
- ***Opportunities***
  - Professional Upgrade Courses
- ***Threads***
  - Hostile managers' collegiate body
  - Hostile organizations board

Training to Improve KMM

In PENTHA model approach, training for pre-service Leaders, as well in-service, is to apply ***COACHING*** and ***NORMING*** modes. Coaching is a method in which the Trainer actively supports the Leader motivating, analyzing performance, providing feedback, reflection on assignment, stimulating discussion about the method adopted. Norming is a method to optimize high performing, effective behaviors

The training purpose is to empower work, purposes and future development of the organizations, according to professionality, vision and mission of the role.

5TH KNOWLEDGE MANAGEMENT WITH FEELING (KMF)

As Orientism component, it focuses on the learning "***TO KNOW MOVING (TOWARDS)***", by Leader to activate a change of perspective or action. It means to activate a general personalized *tutoring plan*, concerning how to support Leaders in their personal choosing process.

SWOT Analysis

- ***Strenght points***
  - emotional intelligences

- ***Weakness points***
    - difficulty to manage stress
    - weak attitude with subordinates
- ***Opportunities***
    - balanced human resource support
- ***Threads***
    - strong hostility between managers or with other Leaders

In operational meaning, emotional intelligence is a set of soft skill for a Leader. It consists in:

- ***Personal Emotional intelligence*** is the ability to manage oneself, own inner environment, autonomy, self-critique, assumption of responsibility (*internal ecology*)
- ***Social Emotional intelligence*** is the ability to manage relationships, the external social environment (*external ecology)*

Successful Leaders are team-builders.

Internal Ecology

Governing Personal Emotional intelligence can support Leader to understand the circumstances and causes that trigger them. Main behavioral features are:

- ***Self-mastery or self-control***
    - Knowledge how to handle anxiety and emotions in a reasonable and profitable manner even under stress, learning to temper reactions
    - attitude of tolerance and understanding of opinions other than theirs (interest in understanding them)
- ***From egocentrism to social self***
    - Availability to listen and understand the other (active listening), to identify oneself in the other (empathy, tuning), to appreciate the contribution of others, enhancing them
    - Knowledge how to think as a community, as well as individuality
- ***Inner strength***
    - Self-esteem, but not arrogance
    - Capacity and availability for self-learning
    - Assertiveness (affirmation of self and ideas in constructive terms)
    - Enthusiasm (the pleasure of working and solving problems)
    - Self-motivation, overcoming difficulties

- *Voltage to the result*
    - Mental orientation towards the production of concrete results and solutions, together with others
    - Professional discipline

External Ecology

Governing Social Emotional intelligence can support Leader to relate positively to others and to interact constructively with them. Main behavioral features are:

- *Understanding the other*
    - Interpersonal communication skills in public speaking (verbal and para-verbal)
- *Authority and leadership*
    - Ability to influence others' behavior by making them useful to the organization and to task
    - Knowledge how to manage with authority and flexibility rather than by virtue of the authority conferred
    - Actively act as Coach of your team of coadjutor, co-workers and subordinates
    - Knowledge how to motivate your collaborators, to delegate when appropriate, possible and without prejudice to security; to decide to risk (under calculated risk conditions)
- *Pushing to continuous improvement*
    - Attitude to solving problems rather than looking for faults
    - Monitoring the progress of projects, studies, interventions and capitalize on acquired know-how
    - Attitude to support innovative ideas and improvement
    - Knowledge how to promote lifelong learning, self-learning
    - Skills immediate positive and, if necessary, negative feedback on the performance
- *Driven to the Institution's result and image*
    - Knowledge how to dress up the institutional roles (manager, member of a group, public representative of an institution, institutional leader, spontaneous, subordinate, participating in a working group etc.)
    - Knowledge how to get the same from your co-workers and coadjutors
    - Ability to promote the image of the institution both inside and outside
    - Availability to offer the best of your skills and commitment

Training to Improve KMF

In PENTHA model approach, training for Leaders is to apply ***COACHING*** (see KMM training), ***NARRATING*** (see KMC/h training), ***FADING*** and ***REFLECTING*** modes. Fading is a method for adjusting and adapting the learning path according to the achievements of the Learner until the proof of his positive capability in full autonomy. Reflecting is a method which pushes to compare own difficulties with an Expert / Tutor and encourages them to perform pull actions. Reflection is the vehicle for critical analysis, problem solving, synthesizing of opposing ideas, evaluation, identifying patterns and creating meanings. The training purpose is to empower spirit of service and love in the role, according to own vocation, passion for the mission.

Computational Thinking

There are four key techniques (cornerstones) for computational thinking:

- ***decomposition***: breaking down a problem or a complex system into smaller and more manageable parts;
- ***recognition of patterns:*** search for similarities between and within the problems;
- ***abstraction:*** focusing only on important information, ignoring irrelevant details;
- ***algorithms:*** development of a gradual solution to the problem or of the rules to follow to solve the problem.

Each cornerstone is as important as the others. They are like table legs: if one is missing, the table will likely tip over. The correct application of all four techniques will help when programming a computer.

Computational thinking in practice and a complex problem is one that, at first glance, we do not know how to solve easily.

Computational thinking involves taking that complex problem and breaking it down into a series of smaller, more manageable problems (decomposition). Each of these smaller problems can then be examined individually, considering how similar problems have been previously solved (pattern recognition) and focusing only on the important details, ignoring irrelevant information (abstraction). Subsequently, simple steps or rules can be designed to solve each of the smaller problems (algorithms).

Finally, these simple steps or rules are used to program a computer to help solve the complex problem in the best way.

Thinking computationally is not programming. It doesn't even mean thinking like a computer, as computers don't think and can't think. Simply put, programming tells a computer what to do and how to do it. Computational thinking allows you to understand exactly what to tell the computer to do.

An example might occur when playing a video game. Depending on the game, to complete a level we should know: which objects to collect, how to collect them and how much time we have where the exit is and the best path to reach it in the shortest possible time what kind of "enemies" there are and challenges.

From these details a strategy can be devised to complete the level in the most efficient way.

6TH KNOWLEDGE MANAGEMENT WITH WILL (KMW)

As Orientism component, it focuses on the learning "***TO KNOW EXPERIENCING***", process that the Leader needs to activate and control. It means both *historical, comparative* and *simulation plans,* about analyzing, using own, and other's previous experiences, comparing transversal situations and conditions, to make decisions, using also virtual simulation tools.

SWOT Analysis

- ***Strenght points***
    - scientific analysis of the decisional process
- ***Weakness points***
    - mistakes of leadership
    - not adeguate or not updated competences
- ***Opportunities***
    - tools for reflecting activities or repository or decisional simulation
    - sharing of ideas between peer
- ***Threads***
    - low or not adeguate human resources

Training to Improve KMW

In PENTHA model approach, training for Leaders is to apply ***FADING*** (see KMF training) and ***PERFORMING*** (see KMC/m training) modes. The training purpose is to empower factual actions, spirit of service and soft skills, according to own passion, vocation and professionality.

7TH KNOWLEDGE MANAGEMENT WITH UNDERSTANDING (KMU)

As Orientism component, it focuses on the learning *"TO KNOW ENABLING"*, that the Leader needs to improve. It means to activate *tutoring*, as well *management* and *social plan*s.

SWOT Analysis

- *Strenght points*
  - Sharing project and organizational design guidelines by verifying their financial compatibility
  - Acquisition of information on the organizing work and security system (structural, hardware/software, web)
- *Weakness points*
  - Timing difficult to reconcile with deadlines of planned commitments
  - Progressive reduction of support activity
  - Difficulty in immediate intervention
  - Lack of timely responses
  - Little knowledge and attention to organizations issues by managers
- *Opportunities*
  - Comparison between peers (Principals) of existing issues and regulatory clarifications
  - Use of regional expertise to solve internal issues
  - Creation of internal / external networks
  - Possibility of establishing partnerships, protocols of understanding, partnerships with private and business associations
  - Quality controls as feedback to be analyzed
- *Threads*
  - Little dialogue with families
  - lack of convergence in educational goals between partners
  - Too much commitment to Leader involves high risk of errors

In assessing the activity of the organizations principals, according to the Italian Educational Reform, the analysis takes into account the socio-economic context in which they work and the results of the processes activated to achieve the goals defined by the organizations within the purpose of the offer training.

Specifically, the evaluation is carried out through indicators identified taking into account the executive's contribution to the pursuit of the results for the improvement of the organizations service provided for in the self-assessment report and the general criteria laid down.

Example of categories of indicators: direction and organization of the organizations institution; internal and external relations; innovation and development; valuing human resources and managing financial and instrumental resources available.

Training to Improve KMU

In PENTHA model approach, training for Leaders is to apply ***COACHING*** (see KMM training), ***NARRATING*** (see KMC/h training), ***FADING*** (see KMF training), ***NARRATING*** (see KMC/h training), ***REFLECTING*** (see KMF training), ***SCAFFOLDING*** (see KMC/m training), and, in addition, ***EXPLORING*** modes. Specifically, Exploring is a method, which forces the Principal to solve problems with new or alternative solutions. The construction of knowledge occurs through the observation and the transformation of experience. The training purpose is to empower awareness of the possible developments and soft skills, according to vision of the future, own vocation and professionality.

8TH KNOWLEDGE MANAGEMENT WITH PERSONALIZATION (KMP)

As Orientism component, it focuses on the learning "***TO KNOW CHANGING***" by the Leader. It means to activate a *creating plan* to improve new dynamics, relations and balance.

SWOT Analysis

- ***Strenght points***
    - Personal co-ordination with regard to the process of change
- ***Weakness points***
    - Fear of change
    - Tendency to individualism
- ***Opportunities***
    - Research new relationships with the work environment
- ***Threads***
    - Issues related to pre-adolescence and adolescence
    - Resistance to change by managers and / or families

Training to Improve KMP

In PENTHA model approach, training for Leaders is to apply both ***SCAFFOLDING*** (see KMC/m training) and ***STORMING*** modes. Storming is a method concerning effective communication, conflict resolution. The training purpose is to empower awareness of role of any involved agent, related to the goal, according to vision of the future, own vocation and own mission.

9TH KNOWLEDGE MANAGEMENT WITH AVAILABILITY (KMA)

As Orientism component, it focuses on the learning "***TO KNOW OPENING***" towards people, environment, prospective and perspectives, interpretations by the Leader. It's part of both *creative* and *social plans*, already described.

SWOT Analysis

- ***Strenght points***
  - Vertical curriculum: educational continuity between organizations orders can lay the foundation for successful training, builds the educational alliance of a territory
- ***Weakness points***
  - Not in-depth knowledge of other areas of organizations, both from an organizational point of view and from the point of view of acquiring knowledge skills and competences
- ***Opportunities***
  - Experimenting with cross-design projects on different stakeholders by exploiting the different experiences and professionalism in the different sectors
  - Active involvement of agents
  - Opening and dialogue with families
- ***Threads***
  - In the interest of continuity of methods and strategies in a vertical curriculum the organizations may lose the specificity of its field

Training to Improve KMA

In PENTHA model approach, training for Leaders is to apply ***FADING*** (see KMF training) and ***STORMING*** (see KMP training) modes. The training purpose is to empower awareness of role of any involved agent, with energy and spirit of service, according to vision of the future, own vocation and passion.

10TH KNOWLEDGE MANAGEMENT WITH SYNERGY (KMS)

As Orientism component, it focuses on the learning "***TO KNOW DEVOTING***" towards work, goals, relationships by the Leader. It's on the base of *all plans*, specifically it's part of *social plan*.

SWOT Analysis

- ***Strenght points***
    - Resistance to a very complex job for passion and strong vision of own administration
- ***Weakness points***
    - No tuning with College of Managers
- ***Opportunities***
    - Financial and operational tools by Government to be able to accomplish innovation
- ***Threads***
    - Too fast changes in legislation, sometimes contradictory

Training to Improve KMS

In PENTHA model approach, training for Leaders is to apply ***FADING*** mode (see KMF), especially adjusting and adapting the decision path according to the achievements of the Leader, until the proof of a synergetic rhythm of work and positive feedbacks.

CONCLUSION

The present contribution intends to be a theoretical contribution for developers in the filed of Artificial Intelligence, as well as a practical guideline for Leaders, as decision makers, to manage tasks towards goal, and to understand what elements are useful to own personal and professional development to optimize performances.

The proposed approach interprets the fluid nature of the decision making process, looking at knowledge and knowledge activities as dynamic, adaptive and self-regulative, based not only on well-known explicit curricular goals, but also on unpredictable interactions and relationships between players. The knowledge process is emerging in human and biological, social and cultural environments. This process is oriented orient on the internal "diversity", the variety of backgrounds, interests, knowledge, skills and the whole personality of the trainees, individually or within in a work team.

Following step is a new concept of e-environment as a Global contributing Knowledge Cells (KC) Technology.

The project describes an advanced, intelligent educational learning, simulation, and exercise environment, assisted by an artificial intelligence (AI) skeleton framework, driven and developed based on latest state-of-the-art pedagogical, didactical and Knowledge Management concepts and paradigms.

REFERENCES

Bandler, R. (1996). DHE. Meta Pubns.

Bandler, R., Robert, A., & Fitzpatrick, O. (2013). *An Introduction To Nlp*. Harper Collins Publishers.

Black & Moudon. (1994). *The Managerial Grid*. Gulf Publishing.

Cheong, R. K. F., & Tsui, E. (2010). The Roles and Values of Personal Knowledge Management: An exploratory study. *VINE Journal of Information and Knowledge Management Systems*, *40*(2), 204–227.

Cheong, R.K.F., & Tsui, E. (2011). From Skills and Competencies to Outcome-based Collaborative Work: Tracking a Decade's Development of Personal Knowledge Management (PKM) Models. *Knowledge and Process Management. The Journal of Corporate transformation, 18*(3), 175-193.

dall'Acqua, L., & Santo, M. (2012). New interpretative paradigms to understand and manage the unpredictability of the dynamics of learning in a complex multi-user knowledge environment. *Proceedings of Artificial Intelligence Workshop 2012 on Collaboration and Intelligence in Blended Learning*.

dall'Acqua, L., & Santo, M. (2014). Orientism, the basic pedagogical approach of PENTHA ID Model vs. 2, to manage decisions in unpredictability conditions. *Proceedings of World Congress on Engineering and Computer Science WCECS2014*, 316-21.

dall'Acqua, L. (2014). *Needs and strategies of KM in a multi-user environment: PENTHA Model view and analysis*. In Jekpot KM19, Milano, Italy.

dall'Acqua, L. (2018). *Improving Business Performance Through Effective Managerial Training Initiatives*. IGI Global.

Hersey & Blanchard. (1982). *Management of Organizational Behavior: Leading Human Resource*. Prentice-Hall.

KEY TERMS AND DEFINITIONS

Computational Thinking (CT): A problem-solving process that includes a number of characteristics and dispositions. CT is essential to the development of computer applications, but it can also be used to support problem solving across all disciplines, including the humanities, math, and science.

DHE: Concentrates on creating powerful states and mental tools so a person naturally moves in the direction they want to go. DHE training consists of learning to create sounds, sights, and internal experience in a way that naturally propels you to be the kind of person you want to be.

Emotional Intelligence: Personal emotional intelligence is the ability to manage oneself, own inner environment, autonomy, self-critique, assumption of responsibility (internal ecology); social emotional intelligence is the ability to manage relationships, the external social environment (external ecology).

Leadership: Is the process of enlisting and guiding the talents and energies of educational community toward achieving common educational aims.

Middle Management: In organizations, a team of managers, responsible for implementing the leader policy.

Orientism: Is a new multi-dimensional KM approach to improve the people's ability to manage decisions and own change of perspective, according to natural, social, artificial environments, in personalized multi-user dynamic, assigned value to multiple reference points and multi-interpreting paradigms.

PENTHA Model: Is an instructional/knowledge design model focused on designing the KM environment, defining the (didactical) rules for an intelligent tutoring system, facilitating person's change and enhancement.

Chapter 2
Optimizing Supply Chains Through System Dynamics Modelling and Simulation:
Lessons From the Navy

Pedro B. Agua
https://orcid.org/0000-0003-1886-9938
CINAV, Portuguese Naval Academy, Portugal

Anacleto C. Correia
https://orcid.org/0000-0002-7248-4310
CINAV, Portuguese Naval Academy, Portugal

Armindo Frias
https://orcid.org/0000-0003-1298-4273
CINAV, Portuguese Naval Academy, Portugal & ISEG, Universidade de Lisboa, Portugal

ABSTRACT

In critical activities and organizations, decision making in the face of complexity has been a growing normal. Complexity troubles humans due to cognitive limitations. Moreover, humans are merely able to understand cause-and-effect relationships that are close in time and space, not the paradigm of many complex socio-technical systems. Decision-making processes shall rely on models that help harness a problem´s associated complexity – among them the dynamics of supply chains. Models typically fall into two broad categories: mental and formal models. Supply chains are complex systems, which may exhibit complex behaviour patterns. Decisions and policies within organizational systems are the causes of many problems, among

DOI: 10.4018/978-1-7998-7126-2.ch002

them undesirable oscillations and other problematic patterns of the parameters of interest. A system is a grouping of parts that work together for a purpose. Hence, the systems dynamics methodology is an adequate approach to deal with fuel supply chain management. A model was developed that helps manage marine gasoil supply chains in the context of the navy.

INTRODUCTION

In critical activities, spanning from business to critical organizations operations, as the military ones for example, decision making in face of complexity has been the growing normal. Such complexity poses difficulties for humans due to cognition limitations, as for instance limited memory and processing capacity, and the inability to follow the behaviour of a set of simultaneous varying parameters over time. Moreover, one can only understand cause-and-effect relationships that are close in time and space, which implies a fair inability to understand and follow causal relationships that are not close in time or space. Therefore, and aiming at better decision-making, we do have to rely on formal models to harness the inner complexity of many problem typologies – among them the dynamics of supply chains. Models can be classified into two broad categories: mental and formal. The former ones are non-shared models and cannot be discussed as they are internal to our minds. The later ones are discussable because being formal they become explicit and can be shared under several formats.

From the several approaches to model complex systems, the System Dynamics methodology, developed at Massachusetts Institute of Technology during the 1950s seems quite suitable to understand and help in the management of complex systems. A System is a grouping of parts that work together for a purpose (Forrester, 1968). Also, according to Forrester (1961) our organization´s decisions and policies are the causes of many unintended problems, among them, undesirable oscillations and other problematic patterns affecting the parameters of interest. Hence, the systems view is an adequate approach to deal with supply chain management, a critical issue for some typologies of organizations where the outcomes are strategic, if not at a national level, at an organizational level. Armed forces among other organizations, fall under this category.

It is critical to bring clarity to the fact that one typically divides systems – sometimes systems of systems – in its composing elements, to cope with their inner complexity and to be able to analyse them. But whatever one divides for purposes of analysis, one must integrate again. Oftentimes a system cannot even be understood when split apart. In the same sense the understanding of a heart outside of the body would not allow for a full understanding of such organ.

One of the main thinking revolutions of the 1950s suggests that one cannot fully understand the nature of systems by analysis, statistical or otherwise. Another method is desirable – synthesis. When trying to understand a system the first thing analysts do is to take it apart; when synthesis advocate exactly the opposite. The defenders of deductive approaches may disagree, however all creativity (hence original ideas) come from induction thinking. Deduction is useful, however, to "validate" or refute a hypothesis or set of hypotheses, for a certain time horizon. Analysts even suggest that the product of analysis is knowledge. Knowledge, however, is not always the same as understanding. Knowledge may even lead to the *know how* it works, but not *know why* it works.

When one gives up synthetic thinking for the sake of analysis, one starts missing understandability. Systems thinking, fundamentally grounded on logics, is a blend of analysis and synthesis, leading towards an expansionism instead of reductionism doctrine (Forrester, 1961; Senge, 1990). Such approach suggests that in order to build increasing understanding one has to move towards larger systems. With synthesis, knowledge goes from wholes to larger wholes, not the opposite – from wholes down to parts.

Causality implies determinism, and logic ensures determinism. Aristotle´s first law of logic states that in a logical implication, if one denies the consequences (effects), one must deny the precedence. Hence, it is a fundamental law of logic that nondeterminism cannot ensure causality. When considering supply chains, we face a similar dilemma. From one side it is easier to understand activities in the field at *tactical* level, however there is a continuum from the whole organization, i.e., system, strategy down to tactics and vice-versa, within feedback control influences, where decisions at the top influence what goes into the tactical level and vice-versa (Figure 1).

Figure 1. The three domains overlap and interact with each other as a continuum

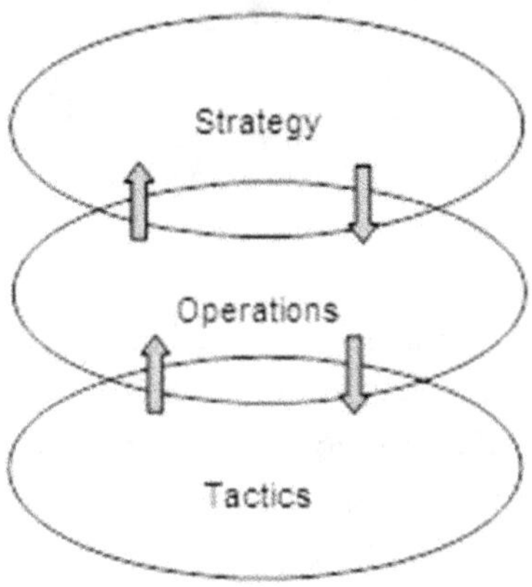

Marine Gasoil (MGO) is one of the most common naval fuels across Navies. Its criticality for naval operations and missions cannot be understated. Many Navies, however, may incur in inefficient management of such resource. Be it because there have been inaccuracies in the whole MGO yearly allocated budget, or because there are oscillations and "bullwhip" phenomena along the MGO supply chain, threatening to disrupt it (Chase & Jacobs, 2017).

Supply chains, as the fuel one, are complex systems, which exhibit complex behaviour over time, which at some stages may resemble chaotic behaviour. By using a scenarios approach, a simulator can predict the due outcomes, uncovering the complex behaviours resulting from the system's underneath structure. A full understanding of the MGO supply chain shall encompass as many variables are necessary, but no more. Here lies the difficulty, more specifically of where to place the systems' frontiers.

Harnessing a fuel supply chain is crucial for critical organizations as would be the case of the Navy, so to ensure minimal risk of fuel supply disruption. Moreover, with a better understandability of the system of interest, one can manage risks and operate with lower levels of employed capital. The first step in trying to understand why a systems' performance is not as expected is to observe key parameters performance or behaviour over time. Why is such performance happening? Where will we end if we do not change course? And how to improve the systems' performance? (Figure 2). This approach brings further understanding about oscillations and potential unintended consequences that may arise from complex adaptative systems and their non-linear behaviours, aka within a MGO supply chain.

Figure 2. Understanding performance over time and objectives
Adapted from Warren (2003).

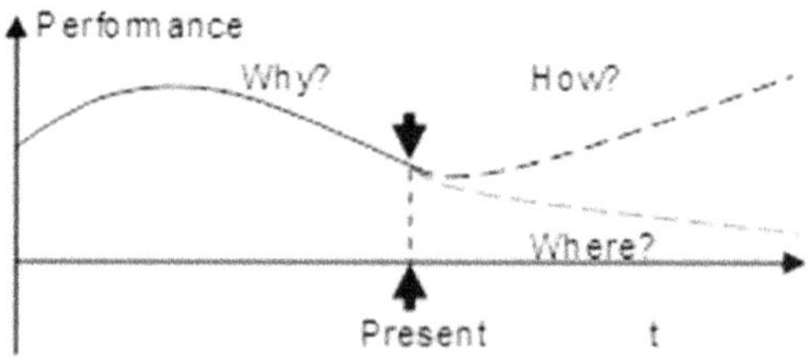

This text is composed of three sections, besides this introduction where the relevance of a MGO supply chain is established. The section on methods provides a brief introduction to the used methodology, which suggests the testing of generated scenarios with a system dynamics simulator. System dynamics modelling is central within the scope of this paper, and the methodology makes use of a "what if…?"

approach, hence providing insight by being a sort of "management flight simulator". The results section presents the outputs from several scenarios' simulations by using the MGO System Dynamics simulator, which by inspection and successive trials build understanding about the MGO supply chain. The conclusion section links this text main argument with supporting evidence from simulation, while at the same time identifying some limiting conditions, while concluding with some derived implications.

The Relevance of Simulation as a Methodology

Since there is no way to evaluate the outcome of a proposed policy without compromising resources, simulation becomes an efficient way to test the plausible effects of choices and operational policies in a safe manner, i.e., without permanently compromising the aforementioned resources. Originally the word *simulate* means, imitate. This suggests that simulating means imitating something, through modelling or a simplified representation of reality regarding the system or problem under study. Simulation models can be of different nature: physical, mathematical, computational, or any combinations thereof. Several simulation types involve physical models. For example, naval engineers have been using wave tanks where a small-scale physical model of a ship, submarine or torpedo is constructed from wood or other appropriate material and placed in such water tank so that its characteristics are studied and its dynamic behaviour understood, before the allocation of considerable amounts of resources for the construction of the real systems is carried out.

Since physical models are relatively expensive and sometimes impractical to test safely (e.g., nuclear facilities), the preference lies with mathematical models, where symbols and equations are used to mathematically model the representative relations of the system of interest (SoI). To perform the simulation using mathematical models, the calculations of the model's equations are performed repeatedly in order to imitate the passage of time. If such calculations were carried out manually, these simulations would have an enormous cost. In the 16th century, years were spent running numerical simulations for the creation of the Nautical Tables, but the great importance of navigation for commerce and naval power showed that it was a well-spent effort. Fortunately, in the last sixty years computer simulation has replaced hand calculations.

By the time of the World War II, with the decrease in the cost of calculating machines ("adding machines"), there was an increase in the use of simulation techniques. As a result of such usage, several other systems emerged, like the Radar, torpedo launching systems, artillery systems and their associated fire control systems. But the cost associated with the simulation was still so high that it was essentially restricted to military projects. It was only with the development of the computer industry in the post-WWII period that there was a spread in the use of computer simulation techniques.

Computer simulation is now common in a wide range of applications, both in the physical sciences and in the social sciences as well. For example, much of the knowledge about the behaviour of nuclear reactors in accident situations is due to computer-modelled simulations, since actual exercises in this field could pose high risks to human life and environment. Hence, computer simulations are of great value in learning how to react in crisis situations.

The widespread use of simulation has been shown to be useful in several areas, such as: project management, business growth management, logistics and military operations, environmental impact, and national economic planning. Therefore, simulation is an adequate approach to study the behaviour of complex systems.

Since we are dealing with complex adaptive systems - systems whose consequences are "far apart" in time and space – the System Dynamics methodology proves to be an appropriate technique for modelling and simulating such systems. For whenever there is a need to deal with feedback and non-linearity phenomena, typical in complex adaptive systems, it is critical to capture and represent the dynamic behaviour of such systems over time.

Moreover, there are advantages in using such approach to support the formulation of organizational and operational policies, as one will end in a hypothetical scenario in which one can assess the potential resulting behaviours, discovering potential unintended effects, derived from the inherent dynamic complexity of the system of interest (Axelrod & Cohen, 1999).

Another advantage derives from the fact that it can be used as a "monitoring and control tool" as well, which will allow the progressive validation of the selected policy implementation, which are usually impacted by the evolution of the operational environment parameters.

System Dynamics

The System Dynamics body of knowledge was developed by Jay Forrester in the late 1950s at MIT and has its roots in the control and servomechanisms theory of the 1930s, an engineering field that faced major advances in the course of World War II. Initially, Forrester applied the concepts of control theory to the study of recurrent oscillations and problems in industrial production, with a view to understanding what a better industrial operations management and business policy design could be (Forrester, 1961, 1968).

Later on, there were several other meaningful works that increased the respective body of knowledge (Forrester, 1968). Following these, Forrester published his famous *World Dynamics* study (Forrester, 1971), whose contribution was recognized by the Club of Rome as serious and relevant, as it allowed to better understand and show the behaviour of the "world system". A methodology was thus launched for improved dealing with the dynamic complexity that can emerge from systems. Since then and aided by the proliferation of personal computers, there has been a growing interest in this field of knowledge.

It is recognized that humans have limited processing and memory capacity, resulting in a considerable inability to understand complexity and causal relationships that are not close in space and time (Sterman, 2000). Humans, however, have the unique ability to decide about "where to focus attention" and model the potential interrelationships behind a system's structure. In this way, by cybernetically combining this capability with the processing and memory capabilities of computers, it is possible to develop better understanding of the complexity inherent to the systems around us or of which we are part of.

The used approach combines the capabilities of the human mind with the capacity of current computers. Humans appear, for now, to be better equipped to formulate structures in which separate pieces of information can be framed or structured. But after such related pieces are modelled as a structure, the human mind is ineffective in anticipating the dynamic behaviour over time that arises from some system. At this point, the computer presents itself as an ideal complement, because it can trace the results of the interaction which results from the internal structure of the modelled system.

The computer is thus instructed by the model it is given to simulate, which consists of a set of descriptions indicating how each part of the system acts with each other. In this way it is possible to construct laboratory models of socio-technical systems, which, although simplifying the real-world systems, are more robust ways of understanding the behaviour of systems than the traditional approach based on mental models, typically used as a basis for industrial (and other) policy debate, only supported by knowledge, experience and intuition.

A computer model is based on the assumptions that may be behind the structure of the system. Yet a model is only as good as the theory that supports it. A good computer model is distinguished from a bad model insofar as it can capture more of the essence of the system it presumably represents. Building a computational model requires one to be explicit about the assumptions on which our mental models are based. When assumptions are clearly exposed, they encourage deeper discussions and lead to better selection of the vast amount of fragmented information contained in a mental model, thereby reinforcing rigour and discipline, generally absent from traditional discussions and descriptions.

Based on the assumptions about how the different parts of a complex system interact with each other, the computer traces the behaviour of the system over time, making use of the mathematical relations and following the rules established in the model description. The computer will thus point out unintended side effects, often stemming from supposedly good intentions.

The goal of any theory is to create a representative model of a problem or complexity that allows it to be explained and understood. Therefore, instead of using comparative static models, which seek to explain the behaviour of systems, supported by covariant properties, System Dynamics in turn, requires the attention to be focused on causal mechanisms and not covariation alone. In this way, the fundamental objective of System Dynamics is to understand the structural causes behind the behaviour of a system and resulting outcomes.

Building a Marine Gasoil System Dynamics Simulator

In building a model of a complex system, the selection of information from the actual system is important, but it should not be an end in itself. According to Forrester (1971, p. 17):

... Generally, we are handicapped not so much by shortage of information as by an excess of information from which to choose, not only is there far more information available than it is appropriate to include, but also the information is unstructured. The unrelated fragments of information yield the structure of the model. Formation of the model should be guided by the principles of structure that are common to all dynamic systems...

System Dynamics uses levels and flows as building blocks and key graphics. The diagrams thus constructed are referred to as *level and flow diagrams*, also named *Forrester diagrams*, after their creator.

System Dynamics models are powerful tools to help us understand and leverage feedback interrelationships in complex systems. Such models offer an operational methodology to support decision-making (Forrester, 1961; Sterman, 2000). Decision makers can use the models to test scenarios, changing constant values and decision parameters in the simulator, in a “what-if?” fashion and explore what could happen, under a variety of different assumptions, past and future, making better decisions about the alternatives available.

Among the various alternative simulation scenarios, it is then possible to choose the options that best meet the chosen objectives. These models help quantify the system’s costs and performance gains, so that decision-making is more informed, exploring and verifying responses to alternative hypotheses or policies.

Having a given system, there are techniques that allow for the management of a system in a way that fulfils a certain intended purpose or objective. It is possible to act on the system in a variety of ways. Meadows (2008, p.145) suggests the following action measures, (from the least important to the most important), in order to change or manage the behaviour of a system:

1. Change constants, parameters, numbers
2. Change the size of levels against their flows (buffers)
3. Changing the structure of levels and flows
4. Change the duration of delays (delays) in relation to system change dynamics
5. The strength of the balancing cycles in relation to the impact they intend to have on the system
6. Gain in positive feedback cycles, supporting growth
7. Changing the structure of information flows in the system (who has access to information in the system or not)
8. Change the "system rules" (incentives, penalties, restrictions)
9. Power to add, change, evolve or reorganize (self-organize) the system structure
10. Changing the "objectives" of the system
11. Change the paradigm that frames the system (and its objectives, purposes, structure, standards, delays, and parameters)
12. The power to transcend paradigms

For example, (1) changing the elements of the system typically produces few effects, (2) changing the interconnections between the elements of a system (i.e., its structure) already produces more effect, but (3) it is by changing the ultimate goal of the system that great impacts are achieved.

Another relevant concept when using systems dynamics approaches is the discovery of leverage points - points where action has a great impact on the behaviour of the system. Using only levels and flows as basic units of graphical representation in the construction of the Forrester Diagrams, the system structure consists only of these two entities. The levels (stocks) represent the accumulations (time integrations) in the system and the flows are the variables responsible for the variation of the levels.

A level accumulates the net amount resulting from the action of the flows over time, thereby adding or subtracting its quantity from the level in question. The levels of a system make it possible to fully describe the same system at a given moment in time. In a way, levels represent the "memory" of a system. The levels define the states and the order of the system of interest (Figure 3).

Four distinct elements can be identified when observing a Forrester diagram: (1) state variables (levels), (2) process variables (flows), (3) auxiliary variables (converters) and (4) causal connectors (arrows), (Figure 4). Auxiliary variables are intended to ensure dimensional consistency and structure, providing a more detailed relation of the influences, for example, on a certain decision variable - flow - validating the base architecture. The fundamental law in System Dynamics

Figure 3. Basic elements in a Forrester Diagram

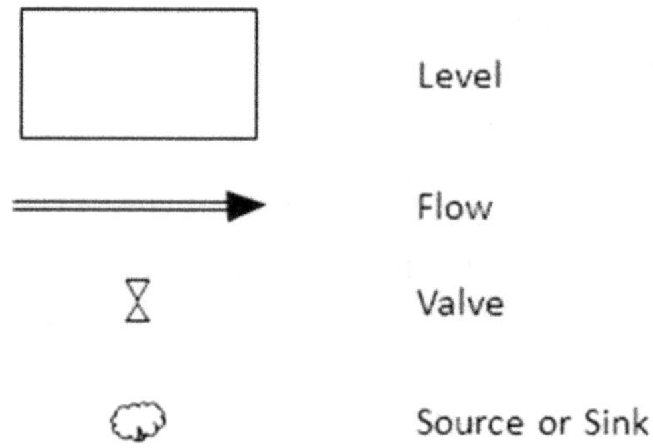

modelling tells us that "a level can only change due to the associated flows". Or in the words of Forrester himself (1971, p. 19): "… *Levels are caused to change only by rates of flow. Conversely, rates depend only on system levels through an information network…*"

Figure 4. Basic forms of structuring a level with the respective flow

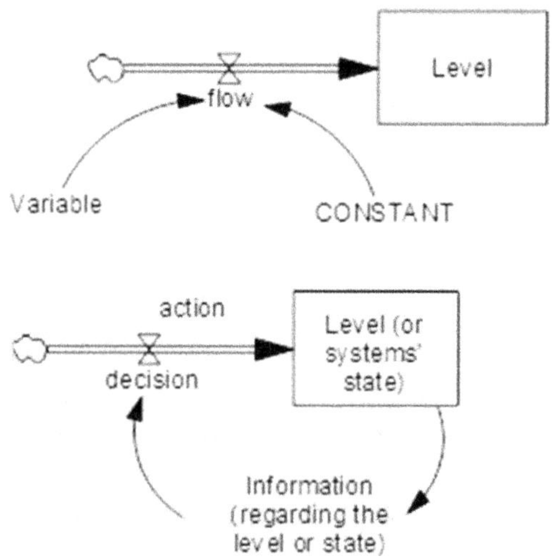

Therefore, all systems which behaviour vary over time can be represented only by levels and flows. These two types of variables are necessary and sufficient to describe a system. Generally, the following arrangement can be made for a level and its associated flows (Figure 5).

Figure 5. Valves are the control mechanisms in System Dynamics. Authors' elaboration

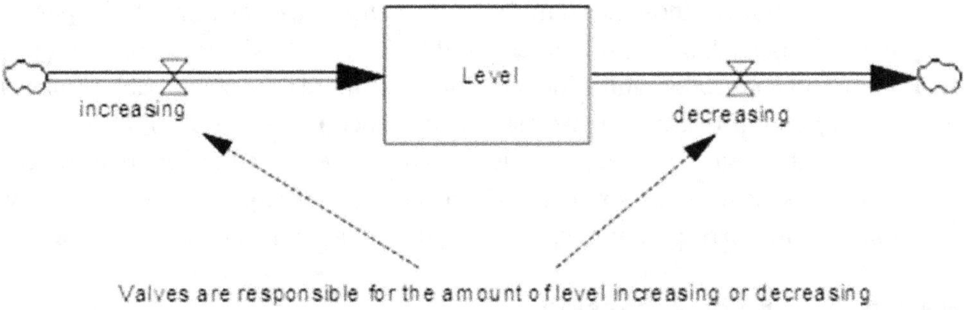

These "valves" are the points in the system where different policies can be selected in order to try to direct the system towards different futures. Mathematically the structure of levels and flows has an underlying integral equation.

$$\boldsymbol{Level}(t) = \int_{t_0}^{t} \left[\boldsymbol{input}(s) - \boldsymbol{output}(s) \right] \cdot ds + \boldsymbol{Level}(t_0)$$

The Dynamic Hypothesis

A central concept in any System Dynamics model is the "dynamic hypothesis", which is closely related with the problem definition and the main impacting parameters (Figure 6). The dynamic hypothesis is an assumption about the dynamic behaviour of a system, subject to stimuli. The very structure of a System Dynamics model, consisting of levels and flows, is in itself a hypothesis or assumption about the causal relations that may be behind the reality or modelled situation.

Figure 6. A strategy process composed of five steps

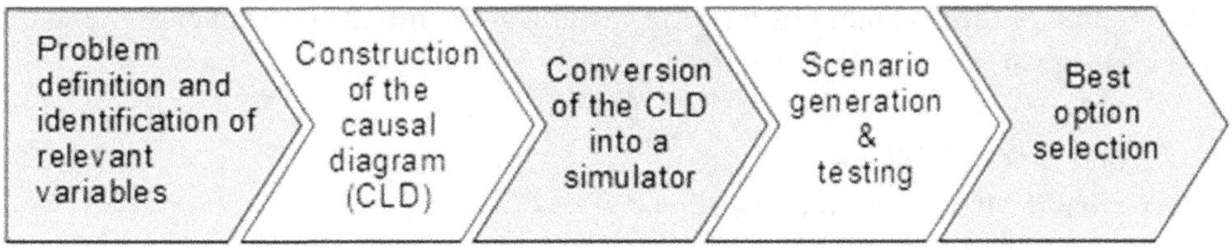

A system modelling approach demands consideration and search for the relationships featuring known systemic effects, such as those resulting from systems archetypes (Senge, 1990), and thus infer possible side effects when adopting concrete solutions or policies for organizational problems. A more robust approach demands however the building of a management simulator, which allows the testing of solutions or scenarios, that is to say, specific policies to be implemented by an organization. Annex A shows the resulting simulator's Forrester Diagram for the case under study, and Annex B the corresponding equations and programming code.

Results: Scenarios Assessing

This section introduces the subject of scenarios and will present some results obtained from simulations. Suffice to say the number of scenarios may as large as reasonable, taking into account the usefulness of them.

An Example from the Navy: The Case of the Marine Gasoil Supply Chain

A study was undertaken as part of a research initiative aimed at building more understandability about the MGO supply chain in naval context (Bituere, 2020). Currently the organization or system under interest relies on traditional operations management practices where the "client" (fleet) issues a request for MGO fuel, which the main supply organization within the navy has to check against the available fuel yearly budget. After such step the naval facilities and logistics support has to coordinate and monitor the fuel supply accordingly. Such process involves human decision and processing along all the needed steps, which are affected by several system parameters and variables.

Such modelling began by identifying the supply chain main variables at play, classified into two main categories: (1) observation variables and, (2) control parameters upon which decisions will affect the performance variables, which makes some of the relevant observation variables. Within the current case the main observation variables of interest are:

- Total Stock of MGO
- Ordered MGO to be received
- Remaining MGO Budget

Figure 7 shows observation variables of interest for the management of the MGO supply chain and the respective influencing root variables or parameters.

Figure 7. Observation variables of interest

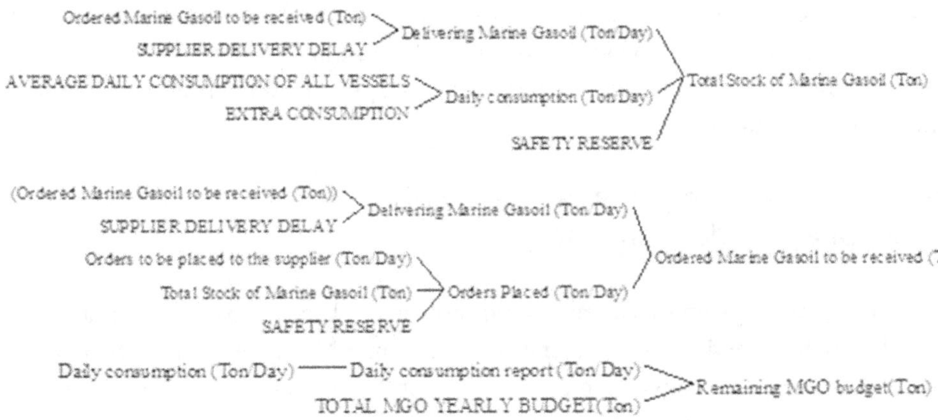

As for the control variables and parameters, aka inputs upon which actions decurrent from decisions will act, the following set is of consideration:

- Adjustment of Stock against remaining MGO Budget
- Orders to be placed to the supplier
- Adjustment of Stocks against consumption

Both types are visible in the Forrester diagram presented in Annex A. On the topside of such diagram, it is made visible the chain of supply, from left to right, where the actual MGO is supposed to flow. While on the bottom side of the diagram represents the information or control network, within which decisions will dictate the flow of MGO in the actual supply chain.

Several parameters represented as capital letters illustrate set-ups of parameters representing time delays to execute some actions within the system, as for example: Time to read storage levels; Admin processing time; Safety reserve; Supplier delivery delay.

The initiating values for all the parameters are realistic ones and can be tested for any scenario. It is useful to initiate a system dynamics simulator under steady state conditions on its parameters, and then introduce "perturbations", according to several selected scenarios. Then, upon simulation, and observation of the behaviour variables will be exhibiting over time, better decisions can be made regarding the adequateness of the assumptions in order to achieve the intended goals – most of these will be a stable fuel supply together with minimizing the risks of fuel disruption while optimizing employed capital in fuel reserves. Once the system dynamics simulator is mimicking reality within an adequate accuracy range, it can be used to test several different scenarios, which in essence is a way to test assumptions and decisions in driving the model behaviour, and draw lessons for the real world. For example, by means of using a scenario where several digitalization of processes occurs, it is possible to reduce delays within the supply chain system, and consequently reduce undesirable oscillations.

Testing "What-if Scenarios" as a Way of Supply Chain Understandability

Having a working systems dynamics simulator, several scenarios can be conceived, according to the intended goals. Scenario generation is a body of knowledge in itself, however some basic process encompassing several steps can be useful in designing such scenarios (Schwartz, 1991, p.226):

1. Identify focal issue or decision
2. Key forces in the local environment
3. Driving forces
4. Rank by importance and uncertainty
5. Selecting scenario logics
6. Fleshing out the scenarios
7. Implications
8. Selection of leading indicators and signposts

Having the above eight steps under consideration, several scenarios can de designed to be tested with the system dynamics simulator, within a "What-if" approach. For the Marine Gasoil case under the conducted study, several different scenarios were designed. The following ones are a sample for exemplification purposes.

Scenario 1: Base Case Scenario

This scenario allows the verification of the behaviour of the MGO supply chain considering average consumptions and without abnormal delays along the supply chains. With this, the fleet send their needs through transfer orders to the Directorate of Supply using an online platform, and such request are forwarded to the fuel supplier. This process can take three days for the requested fuel to reach the storage tanks of the respective vessel. It is stated that this policy would be ideal because it would not allow the build-up of accumulated stocks, and hence a reduction in the amount of employed capital. The values for the simulation input parameters are as follows (Table 1):

Table 1. Base case scenario

AVERAGE DAILY CONSUMPTION OF ALL VESSELS	80 Ton/Day
SAFETY RESERVE	300 Ton
SUPPLIER DELIVERY DELAY	3 Days
ADMIN PROCESSING TIME	2 Days
TIME TO READ STORAGE LEVELS	1.5 Days
TIME FOR ADJUSTMENT OF THE DESIRED QUANTITY	3 Days

Scenario 2: Oil Crisis Scenario

In this scenario, it is considered by hypothesis that there is an oil crisis and fuel supply orders take many days to be fulfilled. This is a critical situation and the Navy must be prepared to respond to a problem of this nature during peace or wartime. In addition to the delay in fuel supply, there is also included in the defining parameters, a one-off extra consumption above averages every 2 months to test the supply chain behaviour. The values of the simulation input parameters are as follows (Table 2):

Table 2. Oil crisis scenario

AVERAGE DAILY CONSUMPTION OF ALL VESSELS	80 Ton/Day
SAFETY RESERVE	300 Ton
SUPPLIER DELIVERY DELAY	20 Days
ADMIN PROCESSING TIME	2 Days
TIME TO READ STORAGE LEVELS	1.5 Days
TIME FOR ADJUSTMENT OF THE DESIRED QUANTITY	3 Days
EXTRA FUEL CONSUMPTION	160 Ton

Scenario 3: Process Digitalization Scenario

In this scenario, business process management considerations are applied to the testing scenario in order to understand how the supply chain performs. This is a scenario in which the digitization of processes is the main focus, and where the system works perfectly without delays in orders and even less during supplies. It is assumed that there is an interaction between the customer and the supplier through an online real-time platform, and the orders are made and automatically met within the established deadlines without causing fluctuations in the supply chain. The values for the simulation input parameters are shown in Table 3.

Table 3. Process Digitalization scenario

AVERAGE DAILY CONSUMPTION OF ALL VESSELS	80 Ton/Day
SAFETY RESERVE	150 Ton
SUPPLIER DELIVERY DELAY	1 Day
ADMIN PROCESSING TIME	1 Day
TIME TO READ STORAGE LEVELS	1 Day
TIME FOR ADJUSTMENT OF THE DESIRED QUANTITY	1 Day
EXTRA FUEL CONSUMPTION	40 Ton, every 2 months

Several other scenarios were and can be designed and tested within this case study. The following figures are outcomes from simulation experiments and illustrate the behaviours over time of some of the variables of interest. By comparing multiple graphs simultaneously, the simulation methodology allows for understandability of the systems as well as better decision-making support, even for new project initiatives aimed at improving the MGO supply chain technological infrastructure.

Figure 8. Simulation curves for the three scenarios. (a) Base case scenario, (b) Oil Crisis scenario; (c) Process digitalization scenario

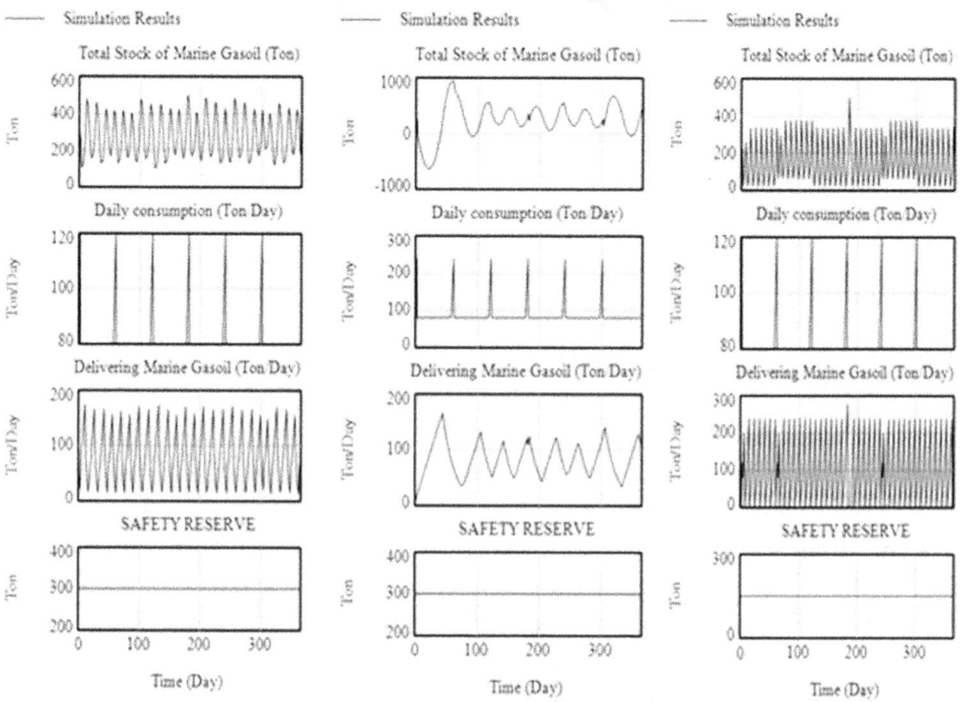

These simulation results illustrate the usefulness of the system dynamics modelling and simulation in order to better understand and manage supply chains, which are in essence complex systems on their own. For example, with the hypothetical introduction of process digitalization (Scenario 3) it is possible to operate with lower fuel "Safety reserves", which besides optimising employed capital, also minimizes the risk of fuel supply disruption. Actually, "disruptive conditions" testing in extreme scenarios is a way of understanding how robust a system is. Moreover, and for each scenario, it is possible to optimise of "fine tune" the selected solutions by means of sensitivity analysis, and comparing simulation results within the same graph.

CONCLUSION

Marine Gasoil is one of the most common fuels for naval vessels, and as such a strategic resource to be managed. It shall be noted that despite the introduced model being applied to a naval context, similar principles and lessons are applicable for many commercial world maritime supply chains, bringing clarity and improved decision-making for businesses that rely on such supply chains. Organizational decisions and policies are often the causes of many problems, among them undesirable oscillations and other problematic patterns of the parameters of interest. A model was introduced to help manage and optimize the MGO supply chain in the context of the Navy, which helps reduce unintended effects as for example the Bullwhip effect. The Bullwhip effect among other oscillation patterns that may arise across a supply chain has its origins rooted on the complexity inherent to many typologies of systems. Such complexity poses difficulties for humans due to cognitive limitations, like limited memory and processing capacity, and the inability to simultaneous follow a set of varying parameters.

The System Dynamics methodology, was chosen to model a fuel supply chain, and a model was developed to help manage and optimize the MGO supply chain for the Navy. It starts by identifying the main relevant variables; then testing several alternative scenarios, mimicking potential fuel disruptive situations, enabling better understanding and learning about such systems' behaviour, which will ultimately help improve the management and operation of such fuel supply chains. Moreover, by reducing oscillations across supply chains, it is possible to increase their performance and minimise employed capital; hence a direct financial benefit.

The main contributions of this paper are of confirmatory nature, as they show understandability about complexity arising from supply chains:

By reducing delays between some of the activities along the processes, the simulator clearly shows a reduction in the magnitude of the resulting systems oscillations, hence ending at a more stable system behaviour.

By being able to operate a more stable system, an organization can reduce its "buffers" or reserves of MGO, optimizing financial management and employed capital over time.

By using a MGO supply chain system dynamics simulator it is possible to test scenarios characterized by extreme conditions under which the fuel supply chain may disrupt.

The simulator also helps in understanding the frequency of occurrence of some non-planned typologies of mission, as for instance: search and rescue missions; humanitarian crises demanding naval relief; conflict escalation; monitoring of fishery areas. With such understanding it would be possible to modify the dynamics model in order to change its parameters according to the distinct probabilities of each scenario.

The system dynamics model can be further tunned in order to account for any major strategic or technological impact as would be the case of digital transformation and the Internet of Things (IoT) where near real time data can help further improve the supply chain system, by means of: less human intervention in the field for purposes of collecting data; automatic order placement; order delivery monitoring and control; payment deadlines monitoring; or optimal economic order placement.

Taken together, these findings suggest that by using a system dynamics simulator, critical organizations as the navy, can operate with less risk of fuel disruptions, while at the same time optimizing employed capital in fuel reserves. Some limitations however need to be pointed out, and concern the need for experimenting with additional scenarios, contemplating high demanding and "war time" types of scenarios and the associated time demands for fuel.

The usefulness of a model in supporting decision-making is of the essence in "validating" such model, however, any model gets fully calibrated only by mimicking reality over a reasonable period of time, which requires the gathering and comparison of field data over the same period of time. The developed systems dynamics model may help improve fuel supply chain management. However, any new order of things will meet with people needs and resistance to change. Hence attention shall be taken on human factors that may affect such implementation (Kotter, 2012). A process can be followed to facilitate change management within organizations. A similar process can be used to management change that cross organizational borders as would be the case of a supply chain where several organizations are involved. The latest case demands special care for cooperation among the involved organizations taking into account the respective organizational cultures.

REFERENCES

Bituere, J. (2020). *Estudo sobre a Otimização da Cadeia de Abastecimento de "Marine Gasoil" na Marinha Portuguesa* [MSc Dissertation]. Portuguese Naval Academy. (in Portuguese)

Chase, R. & Jacobs, F.R. (2017). *Operations and Supply Chain Management.* New York: McGraw-Hill Ed.

Forrester, J. W. (1961). *Industrial Dynamics*. Pegasus Communications.

Forrester, J. W. (1968). *Principles of Systems*. Pegasus Communications.

Forrester, J. W. (1971). *World Dynamics*. Wright-Allen Press.

Kotter, J. P. (2012). *Leading Change*. Harvard Business School Press.

Meadows, D. H. (2008). *Thinking in Systems: A Primer*. Chelsea Green Publishing Co.

Schwartz, P. (1991). The Art of The Long View. New York: Doubleday Ed.

Senge, P. (1990). *The fifth discipline*. Doubleday.

Sterman, J. (2000). *Business dynamics: Systems thinking and modeling for a complex world*. Irwin McGraw-Hill.

Sterman, J. (2001, Summer). System Dynamics Modeling: Tools for learning in a complex world. *California Management Review*, *43*(4).

Sterman, J. (2002, Winter). All models are wrong: Reflections on becoming a systems scientist. *System Dynamics Review*, *18*(4), 501–531. doi:10.1002dr.261

Warren, K. (2003). *The Critical Path: Building strategic performance over time.* Volapress.

ADDITIONAL READING

Badiru, A. B. (2014). *Handbook of Industrial and Systems Engineering*. CRC Press.

Blanchard, B. (2003). Logistics Engineering & Management (6th ed.). New York: Pearson.

Blanchard, B., & Fabrycky, W. J. (1998). *Systems Engineering and Analysis*. Prentice hall.

Ford, D. N., & McCormack, D. E. M. (2000). Effects of Time Scale Focus on System Understanding in Decision Support Systems. *Simulation & Gaming*, *31*(3), 309–330. doi:10.1177/104687810003100301

Hanneman, R. A. (1988). *Computer-Assisted Theory Building*. Sage Publications.

INCOSE. (2015). Systems Engineering Handbook (4th ed.). Hoboken, NJ: John Wiley & Sons, Inc.

Jones, J. V. (2006). Integrated Logistics Support Handbook. New York: McGraw-Hill Ed.

Ritchie-Dunham, J. L., & Rabbino, H. T. (2001). *Managing from Clarity: Identifying, Aligning and Leveraging Strategic Resources*. John Wiley & Sons, Inc.

Warren, K. (2008). *Strategic Management Dynamics*. John Wiley & Sons.

KEY TERMS AND DEFINITIONS

Bullwhip: The bullwhip effect is a supply chain phenomenon in which demand forecasts yield supply chain inefficiencies. It is related with increasing swings in inventory in response to shifts in consumer demand as one moves further up the supply chain.

Causal: A driving or influencing relationship between two variables, in contrast to pure correlations.

Feedback: When the effect of a causal impact comes back to influence the original cause of that effect.

Flow (Rate): The movement of things between stocks within a system. Flows are changes in stocks (levels) over time. Flows represent activity, in contrast to stocks that represent the state of the system.

Information Delay: A delay that represents the gradual adjustment of perceptions or beliefs, or a gradually delayed impact of some variable.

Marine Gasoil (MGO): Marine fuels that consist exclusively of distillates.

Mental Model: Mental models represent the relationships and assumptions about a system held in a person's mind.

Negative Feedback: Feedback that works to cancel deviations from a goal. In isolation or if dominant negative feedback generates goal-seeking behaviour.

Nonlinear Relationship: A causal relationship between two variables in which the change in the impacted variable is not proportional to the change in the impacting one.

Oscillation: Behaviour exhibited by a second-order or higher-order system in which the stock value moves sinusoidally over time.

Policy: A deliberate set of principles to guide decisions and achieve rational outcomes.

Simulation: The generation of the behaviour of a system with a model of the system structure, either using a mental model or a formal computer model.

Stock and Flow Diagram: A graphic depiction of the stocks and flows in a system and how they are connected.

Stock or Level: An element of a system that accumulates or depletes over time. Stocks are the memory of a system and are only affected by flows.

Structure: The arrangement of and relations between the parts or elements of something complex.

System: A grouping of parts that operate together for a common purpose.

Systems Thinking: The use of conceptual system models and other tools to improve the understanding of how the feedback, delays, and management policies in a system's structure generate the system's behaviour over time. Systems thinking involves seeing interrelationships instead of linear cause-effect chains, and seeking processes of change over time rather than snapshots.

ENDNOTE

1 The model was developed using *Vensim PLE* ® software package.

APPENDIX

Annex A – The MGO Forrester Diagram[1]

Figure 9.

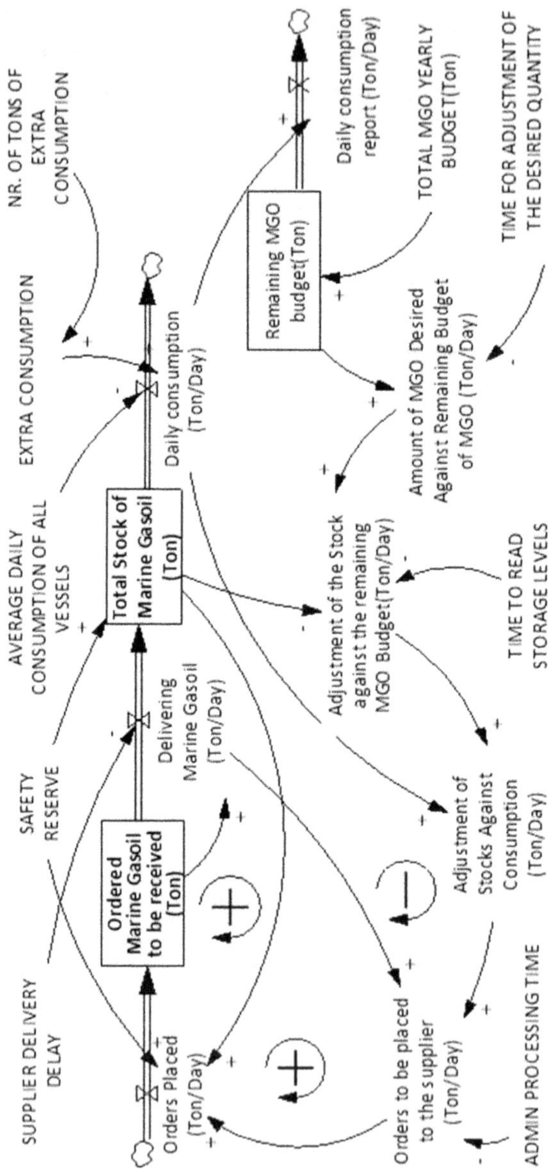

Annex B – Simulator code and equations

(01) "Adjustment of Stocks Against Consumption (Ton/Day)"= IF THEN ELSE (("Adjustment of the Stock against the remaining MGO Budget (Ton/Day)")>="Daily consumption (Ton/Day)","Daily consumption (Ton/Day)", 0)

Units: Ton/Day

This variable assists in the determination of new orders. It has a condition that is: If the Stock Adjustment to the remaining MG Budget is greater or equal to the amount consumed, therefore, this amount consumed, constitutes a new need (an order to be placed). This condition is introduced because, after adjusting the budget and having the information about the remaining budget, one does not use the remainder as a whole to make new acquisitions necessarily, that is, acquisitions depend on immediate needs.

(02) "Adjustment of the Stock against the remaining MGO Budget (Ton/Day)"="Amount of MGO Desired Against Remaining Budget of MGO (Ton/Day)"-("Total Stock of Marine Gasoil (Ton)")/TIME TO READ STORAGE LEVELS

Units: Ton/Day

Stock Adjustment indicates the rate by which one knows what is the level of stock desired.

(03) ADMIN PROCESSING TIME= 1.5

Units: Day

Admin time it takes to verify the quantity of fuel in transit and the consumed, in order to proceed with the process of applications for new quantities (Administrative Decision).

(04) "Amount of MGO Desired Against Remaining Budget of MGO (Ton/Day)"= "Remaining MGO budget (Ton)"/TIME FOR ADJUSTMENT OF THE DESIRED QUANTITY

Units: Ton/Day

The desired stock depends on internal and external variables in the Supply Chain. It may depend on the quantities the supplier deposits into the Navy storage tanks, or the available budget.

(05) AVERAGE DAILY CONSUMPTION OF ALL VESSELS= 80

Units: Ton/Day

The average fuel consumption per day calculated for all ships. It is a variable that represents the amount of Marine Gasoil consumed per day for all ships on mission.

(06) "Daily consumption (Ton/Day)"= AVERAGE DAILY CONSUMPTION OF ALL VESSELS+EXTRA CONSUMPTION

Units: Ton/Day

Fuel outflows associated with average consumption of ships, is a regulator of outflows of tons of fuel. It is possible to know how much fuel is spent for periods of time through this variable. It serves as an indicator of demand for new quantities.

(07) “Daily consumption report (Ton/Day)”= “Daily consumption (Ton/Day)”

Units: Ton/Day

Average consumption per mission performed. At the end of each mission the vessel draws up a map of the quantity consumed and, with this information, new quantities are requested in order to keep the MGO storage tanks levels above 60%.

(08) “Delivering Marine Gasoil (Ton/Day)”= “Ordered Marine Gasoil to be received (Ton)”/SUPPLIER DELIVERY DELAY

Units: Ton/Day

It represents the rate of purchase that is determined by the quantity of orders placed and the average delay in purchase. In this simulator the acquisition delays have a significant influence (they negatively influence because a delay in supply can compromise the mission)

(09) EXTRA CONSUMPTION= “NR. OF TONS OF EXTRA CONSUMPTION”*PULSE TRAIN (1, 1, 60, 360)

Units: Ton/Day

Starting on day 1, and for 1 day, with the monthly frequency (30/60 days), until the end of the commercial year (day 360), will consume periodically more NR. Ton of Extra Consumption.

(10) FINAL TIME = 365

Units: Day

The final time for the simulation.

(11) INITIAL TIME = 0

Units: Day

The initial time for the simulation.

(12) “NR. OF TONS OF EXTRA CONSUMPTION”= 40

Units: Ton/Day

It is the amount of fuel that is consumed on certain days with unplanned missions. Such as Search & Rescue or other emergency services, with the aim of contributing to the safeguarding of marine resources and ensuring safety at sea.

(13) “Orders Placed (Ton/Day)”= IF THEN ELSE ((“Total Stock of Marine Gasoil (Ton)”) <= (SAFETY RESERVE),”Orders to be placed to the supplier (Ton/Day)”, 0)

Units: Ton/Day

Order placed to the supplier. The net order live is obligatorily not negative. If there is a cancellation of a Transfer Request, it must be reported so that it is not included in orders present in the supply chain.

(14) “Ordered Marine Gasoil to be received (Ton)”= INTEG (“Orders Placed (Ton/Day)”-”Delivering Marine Gasoil (Ton/Day)”, 150)

Units: Ton

The amount of fuel ordered; what was requested but not yet received. Notwithstanding the above, it may also include quantities in transit.

(15) “Orders to be placed to the supplier (Ton/Day)”= DELAY FIXED ((“Delivering Marine Gasoil (Ton/Day)”+”Adjustment of Stocks Against Consumption (Ton/Day)”), ADMIN PROCESSING TIME, 0)

Units: Ton/Day

Orders to be placed are proportional to the quantities related to the Stock Adjustment for Consumption and Deliveries of Marine Gasoil considering the Administrative Processing Time.

(16) “Remaining MGO budget (Ton)”= INTEG (-”Daily consumption report (Ton/Day)”, “TOTAL MGO YEARLY BUDGET(Ton)”)

Units: Ton

This is the remainder of the difference between the budget (less the due number of tons) and current needs. In this case it is considered that the quantity consumed should be deducted from the budget, constituting a new need for the supply chain.

(17) SAFETY RESERVE= 120

Units: Ton

Amount of fuel available at the beginning of the economic year (to ensure stocks in the Total Stock), while adjusting the Budget for new purchases and also ensures that there is a safety margin for consumption. This variable will be included in the current year’s budget.

(18) SAVEPER = TIME STEP

Units: Day

The frequency with which output is stored.

(19) SUPPLIER DELIVERY DELAY= 3

Units: Day

This is how long it takes the MGO supplier to supply the fuel. This time represents an operation delay (related to internal factors inherent to the supplier). In short, that’s how long it takes to get the tanks full.

(20) TIME FOR ADJUSTMENT OF THE DESIRED QUANTITY= 3

Units: Day

This is the time it takes to adjust the quantities between the available budget to acquire new quantities and the total stock of Marine Gasoil.

(21) TIME STEP = 1

Units: Day

The time step for the simulation.

(22) TIME TO READ STORAGE LEVELS= 2

Units: Day

This is how long it takes to read the Navy MGO storage tanks and use this information to make decisions in the supply process of ordering new quantities of Marine Gasoil.

(23) "TOTAL MGO YEARLY BUDGET(Ton)"= 33000

Units: Ton

Initial budget in tonnes of fuel according to yearly. At the beginning of each economic year the fleet receives planned information where it is possible to know what quantities of Marine Gasoil it will need in order to fulfil the assigned/planned missions.

(24) "Total Stock of Marine Gasoil (Ton)"= INTEG ("Delivering Marine Gasoil (Ton/Day)"-"Daily consumption (Ton/Day)", SAFETY RESERVE)

Units: Ton

It is the difference of quantities between the accumulated stock of purchases to the consumed stock, i.e., it is the availability of marine gasoil at present. It is from this quantity that quantities to be purchased are determined. In short, one can say that it is the quantity available in marine tanks.

Chapter 3
Risk Models and Managerial Strategies for the Global Supply Chain's Security:
An Analysis of the Management of International Vulnerabilities and Emergencies

Francesco Saverio Zamboni
Alma Mater Studiorum, University of Bologna, Italy

ABSTRACT

This chapter aims to investigate the effectiveness of managerial strategies to address sources of risk and strengthen global supply chain security, with specific attention to disasters deriving from the interaction between extreme natural phenomena and pre-existing socio-economic vulnerabilities. The object of the study is represented by the mutual learning opportunities that can arise from socially-oriented cross-sector collaborations between humanitarian agencies and enterprises. With the ultimate scope of grasping the economic impacts, the business perspective provides several insights to integrate a proactive approach into the corporate strategic planning to ensure a capacity to react with respect to external stimuli. In particular, an in-depth study was carried out dedicated to the regulatory standards that allow the development of this management system. Finally, the private sector's involvement in humanitarian programs is analyzed in order to highlight the benefits in terms of effectiveness of operational performances and the safety of communities' critical infrastructure.

DOI: 10.4018/978-1-7998-7126-2.ch003

INTRODUCTION

In the nowadays society, the importance of global supply chains has experienced a fundamental growth due to the constant increasing flow of physical goods and services. A disruption within any of these flows may lead to far-reaching negative consequences. Therefore, in the last decade, there has been increasing emphasis on managing risk and challenges to improve the overall economic and social security. Following terrorist attacks, for example, firms have been taking multiple steps to ensure safe transit of their goods across international borders. In parallel, natural disasters such as hurricanes, earthquakes, epidemics as well as many other unforeseen events such as product contamination, shortages, border closings, made firms more aware of their vulnerability and weaknesses, and encouraged them to seek strategies to reduce those risk, increasing stability and efficiency along their supply chains.

The current pandemic crisis has underlined several limits of the global interconnection experienced in the contemporary era. The outbreak of the Coronavirus disease had a severe impact on the supply chain of companies. Many of them have not been able to take proactive measures to cope with the stagnant distribution of their products and services, and sharp decline in sales, while others are restructuring their operational and organizational dimensions in view of the post-COVID-19 scenario. In a recent survey, Shigeyuki Ishida claimed that supply chain corporations had a significant impact on business continuity planning (BCP) during disasters (Ishida *et al.*, 2016). As a result, the front line of competition should shift to more comprehensive proactive modelling.

Historical data indicate that the total number of natural disasters has dramatically risen over the last ten years (Tang, 2006). According to Thomas and Kopczak (2005), they are even expected to increase another fivefold over the next 50 years, as ascribable to different factors like global warming, population growth rate, urbanization, residential densification, economic and financial global contingencies, natural resources immoderate use and depletion. Due to these reasons, offering timely and necessary aid to those in need through efficient humanitarian supply chains is a major challenge and logistics acts as a strategic role, as expressed by Trunick (2005). Therefore, the field of humanitarian logistics and supply chain management has recently gained much attention in research. Apart from the frequency and impact of disasters, humanitarian organizations are under an increased pressure of improving their logistics performance.

The chapter aims at analysing the learning opportunities, hidden in the collaboration between private and public in relation to disaster relief operations, to facilitate the exchange of skills and experiences with a view to increasing responsiveness, preparation, and security of the global supply chain leveraging risk mitigation strategies. The research explores the elaboration of proactive approaches able to develop tools and action plans to reduce natural risks at the lowest level.

The field of study relies on the logistics dimension as a mean to stabilize the contiguity in the public and private organizations' strategic planning, reaching a synthesis of the two perspectives through the analysis of a cross-sectors social oriented cooperation. The purpose is to describe an organizational model apt to improve the responsiveness to external environments, developing long-term programs to deal with the increasing of global threats.

BACKGROUND

The term disaster refers to a serious disruption of the functioning of society, causing widespread human, material or environmental losses which exceed the ability of the affected people to cope using only its own resources" (United Nations, 1992). Four types of disaster have been identified in literature (Van Wassenhove, 2006): (i) natural sudden onset, e.g., earthquakes or hurricanes; (ii) man-made, sudden onset, e.g., terroristic attacks; (iii) natural, slow onset, e.g., famine or drought; (iv) man-made, slow onset, e.g., refugees' crises or social and political instabilities. In this context, relief refers to "the emergency food, shelter and services provided in the immediate aftermath of a natural or man-made disaster" (Thomas, 2003) and often entails a foreign intervention into a society with the aim of helping local people (Long and Wood, 1995). As consequence, according to the event, it is possible to subdivide the humanitarian interventions in two categories: disaster relief, meaning the former response efforts in case of natural catastrophes, for instance tsunami; continuous aid work, such as development and assistance programs to fight poverty, famine or social-political crisis (Cozzolino, 2012). In particular, the latter allowed to focus on the management dimension at the strategic and operational level. Indeed, the concept of disaster management describes the set of rules needed to coordinate the activity on the ground using the available resources rationally (Ilha, 2011). Essentially, this method aims at reducing the negative effects resulting from natural disasters. In order to reach this goal, the predisposition and the optimization of competencies and materials it is crucial as well as a developing a flexible and agile systems to manage resources and people. In fact, the goal of humanitarian operations has been defined by Tomasini and Van Wassenhove (2004) as "a successful operation [that] mitigates the urgent needs of a population with a sustainable reduction of their vulnerability in the shortest amount of time and with the least amount of resources".

The last twenty years have seen a growing commitment between the private and non-profit sectors. According to Cozzolino, an important opportunity for aid agencies to learn from companies is, for example, in terms of ability to develop knowledge management systems that make it possible to enhance the experience acquired in the field during emergency operations, which otherwise very often risks being lost (Cozzolino, 2014). Similarly, cross-learning opportunities could provide humanitarian sector with significant insights into how corporations master the process of planning, implementing, and controlling their supply chain operations (Van Wassenhove, 2006). On the other hand, firms have also increased their interest in the humanitarian field. In addition to a pure philanthropic contribution, companies may be interested in reaching certain geographic areas, after a catastrophe, to build new relationships with local governments in order to identify new markets and open new business opportunities in those countries where they are not yet present (Cozzolino, 2014). Companies may also be interested in restoring the continuity of their business after a disaster and in playing an active role in relief operations where their plants, offices, employees, suppliers and customers are located; in this way, they will be able to personally support their disaster-struck business. In addition to these reasons, a trend indicated by Kanter (1999) is concretizing according to which smart companies are approaching the social sector as a learning laboratory, especially in terms of learning potential of "complementary skills" (Oglesby and Burke, 2012). One of this crucial expertise has only recently been identified by some authors (Cozzolino, 2013; Tomasini and Van Wassenhove, 2009) in the "agility" of the supply chain, that is, the capacity of humanitarian logistics to respond effectively and rapidly to unforeseen and serious situations, on the demand side and on the supply side (Lee, 2004). This intuition arises from the observation of the specific experience that the humanitarian sector has developed in the organization of emergency response operations according to the principle of agility.

From an economic point of view, the risk of natural disasters highlights the importance of integrating the emergency management to the firms' strategic planning. In this context, the experiences of international humanitarian agencies could be useful to private sector in order to develop theoretic tools and practical measures to reduce the impact of negative consequences. Indeed, the global supply chains cope with several threats. According to the literature, two typologies have been identified: internal risks, for instance volatility of market demand, uncertainty of delivery times or problems related to the goods' quality and safety. On the other side, it considers as external risks the volatility of monetary exchange rates, terroristic attacks and natural disasters (Kleindorfer and Saad, 2005; Ravindran and Warsing, 2013). The latter represents the focus of the following analysis. Extreme events such as the heavy flooding in China and Thailand of 2011 (EM-DAT, 2011) demonstrate the exposure and the weakness of the global supply chains. In what concern the Thailand case, the disaster has been resulting from the duality of the country environment and the lack of preemptive measures (Ye and Abe, 2012): inadequate urban planification, political instability and deforestation have determined the failure of the response to the crisis. Several industrial areas have been seriously damaged creating obstacles and leading to the paralysis of the production (ibidem). Moreover, it has triggered a ripple effect. As consequence, other productive sites have been involved due to the unavailability of fundamental components previously provided by the affected suppliers. The example reveals how an unpredictable phenomenon could generate far reaching negative repercussion within interconnected realities. Thus, the full awareness of challenges and external threats could be the mean to improve the stability of a supply chain and could also be at core of new opportunities. In the nowadays uncertainty, the risk management could be perceived as an approach which offers a competitive advantage improving the overall performances (Hauser, 2003).

The survey of Harvard Business Review (2011) claimed that over the 67% of the interviewed managers consider the management of this kind of risk as a fundamental matter to the global economy. Indeed, the 89% of their apprehensions is related to natural catastrophes, rather than financial crises (ibidem). As far as the structure of the supply chains expands at global level, becoming potentially vulnerable to a number of negative events, it is perceived as necessary the ability to improve these means and their integration in long term corporate strategies. A recent research of Birkie and Trucco (2020) confirms the hypothesis according to whom business continuity and risk management measures might mitigate the potential impact of this kind of interruptions, demonstrating how the complexity of the supply chain results in a regulatory positive effect insofar it would be affected by a stronger pressure. In other terms, a more articulated system allows to dislocate either the risks or the potential impact (ibidem). This observation suggests an evolution with respect to traditional approaches where the supply chain management refers to simplicity, efficiency by minimalizing superfluous elements. Whether the main goal consists in the maximization of the profits through the implementation of processes allowing the reduction of unmatched market demand and delivery times of goods, it is possible to assume that an organization has a further objective in order to avoid losses, that is the mitigation of disruption risks.

Nevertheless, it is necessary to understand the major mechanism at the base of economic and social vulnerabilities. The international humanitarian community has recognized the importance of a systematic approach taking into consideration the whole society, committing itself to the reduction of disaster risks. The tendency to implement participative programs meant to increase the resources and competencies at the disposal of the individuals, is shared at the global level in order to manage in a sustainable and profitable way the recovery from natural catastrophes. The disaster risk management is a crucial element of the present and the future developing paradigms, considering the growing movement of economic resources towards risky areas (UNDRR, 2019). In this context, political institutions have the role to create incentives and implement adequate measures. If properly managed, the global economic development offers the opportunity to build up the resiliency of the society taking part to initiatives whose aim is to invest in protection and prevention infrastructures at the local level.

GLOBAL SUPPLY CHAINS AND THE TREATH OF EXTREME EVENTS

The modern evolution of the value chains has led many firms towards a model that allows the outsourcing of several internal process to different external suppliers. In this context, a corporate strategy based on the principles of disaster recovery, emergency management and business continuity, seems to be an efficient tool to identify, profile, assess, manage the impact of emerging threats, and thus cope with an high level of uncertainties. Awareness and realism remain two crucial pillars of the management custom, but in order to develop an agile and resilient supply chain it is necessary to adopt a holistic approach to support the decision-making process by integrating risk management's practices. Before discussing its practical applications and implications, it is useful to illustrate the regulatory compliance rules which examine plans, policies, and management procedures of risks by each organization.

Starting from the 2000s, several public and private entities have expanded and developed a set of regulatory standards at the international level in order to provide guidelines and facilitate implementing efficient procedures in different operational areas. Published for the first time in 2009 and updated in 2018, the ISO 31000 norm focuses on risk, providing general principles and recommendations related to potential room for improvement. Although it cannot be used as a certification, the norm supplies useful details to different activities. For instance, its aim is to increase the probabilities of reach the goals, to ameliorate the opportunities and threats' identification, but also to optimize the exploitation of resources to the risk treatment (International Organization for Standardization, 2009). Aware of that, it is possible to claim that an efficient management system finds its ultimate scope in the value creation for the organization; the systemic integration into the decision-making process; the continue monitoring in order to face each incertitude in an inclusive and transparent way (ibidem). The risk mitigation strategies emerging, concentrate themselves on the inevitability of some catastrophes and aim at taking care of consequences that could arise after the occur of a destructive event.

As a matter of fact, the outbreak of cases of A(H1N1) flu in several countries in 2009 foreshadowed a remarkable increase of the pandemic risk, drawing attention of the OMS which monitors the epidemiologic progress (Repubblica, 2009)[1]. Among the most affected countries, the US recorded over one hundred thousand cases of infection and the CDC's National Center for Immunization and Respiratory

Diseases underlined that almost the 40% of national labor force has been infected or housebound to assist a family member (ibidem). This example allows to understand the importance of preparation and elaboration of an action plan to cope with any crisis: in this specific case, the danger arisen from this kind of risk, consisted in the probability that in a short time frame several employers either the company's ones or its commercial partners could be exposed to the virus. The concept of business continuity management refers precisely to the ability of an organization to maintain an acceptable production level even before an adverse event (International Organization for Standardization, 2012). The international reference standard is represented by the ISO 22301:2019 norm which describes the crucial aspects for an updated system. Several experts have underlined how this approach shares fundamental elements at the core of the relief programs intended for communities affected by natural disasters (Herbane et al., 2010). Indeed, this paradigm consists of specific activities such as the operational management of the crisis, and the disaster recovery, meaning the rehabilitation of critical systems and industrial plants (Norman and Jansson, 2004).

The following sections analyzes two cases of study. On one side, it will be highlighted the firms' perspective in the security management. However, it is a partial approach since the interactive nature of risks within the dynamic global context requires a broader strategic direction and an internationally concerted action. Aware of the economic interdependency, the private sector has developed a network of cooperation in order to face the hidden vulnerabilities exploiting one another competencies and expertise. On the other side, the last section will assess the collaboration between humanitarian agencies and private partners in the creation of an emergency supply chain whose aim is to restore the correct behavior of the interested infrastructure.

THE ERICSSON'S MODEL

The Swedish multinational, *Telefonaktiebolaget LM Ericsson* (henceforth as Ericsson), operating in 180 countries, is specialized in the supplier of technologies, communication services and infrastructure in the ICT sector (information and communication technologies). Among the major suppliers at the international level of mobile communication services, in the last twenty years, Ericsson has outsourced several of its assembly and production activities towards a set of companies located worldwide. As several other multinationals adopting this model, Ericsson has experienced a sever increase in its exposition to a wide range of risks and threats, making it necessary an afterthought of the management strategies towards a more effective policy, aims at assuring the business continuity. Currently, the sustainability and corporate responsibility's principles are at the core of the business activity, whose ultimate scope is to create a long-lasting value, pursuing the reduction of risks and vulnerabilities and generating a positive impact on the society (Ericsson Annual Report, 2019). This effort focuses on the improvement of performances and efficiency related to the whole supply chain, guaranteeing the company a great competitiveness at the global level. This analysis has been redacted thanks to the accessible information and open data regularly provided by Ericsson through annual reports. For instance, it has been possible to describe the most significant example which has underlined the urgence of this kind of approach, triggering a process of improvement in the risks' management of the supply chain.

THE ALBUQUERQUE ACCIDENT

The 18 March of 2000, the city of Albuquerque in New Mexico (US) was hit by a heavy storm: the lightnings generated a power surge all over the State. The meteorological event hit one of the Phillips Electronics NV's industrial plants, which deals with the production of cellular components and finds its major buyers in Nokia Corporations and Ericsson. Due to the lack of an emergency system that could compensate the energetic flow, the ventilation system stopped, causing a fire soon after extinguished.[6] The main issue relied in the absence of an alternative source of energy to power the industrial activities. At first the episode has been perceived as neglectable due to its brevity, lasting no more than ten minutes. However, its impact on the Ericsson production chain was devastating. In the spring of 2001, the company declared a potential loss of almost 200 million dollars, due to the interruption of the furniture of radiofrequency chip produced in that industrial plant, albeit the partial compensations provided by the assurance companies (Ibidem). Indeed, it required almost three weeks to re-establish the production and, after six months, the operational level of the Albuquerque plant was exclusively at its 50 percent (Norrman and Jansson, 2004). Since this industrial plant was the exclusive partner producing that fundamental component, Ericsson was not able to sell and distribute one of its key products until the complete recovery of Albuquerque's production level (Ibidem).

The episode demonstrates how the impact of crises situations, even bounded and contained, could result in massive negative consequences, damaging several commercial partners. In the aftermath of the accident, the Swedish multinational company developed a new management approach, able to preemptively estimate and manage risks that could affect its supply chain and rapidly reach decisions in case of unexpected interruption or similar events.

ORGANIZATIONAL STRUCTURE

Thereafter, Ericsson has successfully implemented a set of tools and measures at the internal operational level. In particular, the risk management process is divided into 4 steps: identification, analysis, treatment, monitor and control (Norrman and Jansson, 2004). Additionally, it has been studied and arranged emergency plan providing preventive actions and guidelines to follow in the event of an imminent danger.

Regarding the internal structure of the organization, the Swedish multinational relies on a global management system, known as Ericsson's Group Management System (EGMS), consisting of a number of sections: Business Areas, responsible for developing competitive solutions such as product or services; Market Areas, devoted to the sales activity and to the customer relations; Group Function, deals to support and harmonize different levels and tasks providing a shared platform with aim of guiding the execution of the corporate strategy (Ericsson Annual Report, 2019). Most of all, EGMS relies on the ISO 9001:2015 certification, that is the international reference standard for quality systems management, designed to describe the elements needed to develop a dynamic governance structure able to adapt to evolving expectations and requirements from stakeholders, partners and customers (ibidem). Thus, the decision-making process manages to incorporate the whole business intelligence cycle from the strategic planning to the implementation of goal-oriented actions. Described with the standard ISO 31000:2018, the Enterprise Risk Management (ERM) was adopted by Ericsson and integrated into the EGMS with the purpose of make compatible these practices and measures with the operational execution of the corporate strategy. The ERM provides methodologies to identify, assess, and mitigate risks as well as determine the society's degree of risk appetite and risk tolerance. Following these principles, Ericsson has developed a complex system where each area and sub-unit plays a crucial role guaranteeing transparency and verification of threats at the central level. On top of this configuration, an Executive Team's member is responsible for the management of the principal identified risks and shall coordinate with the Group Risk Management Function, namely the set of managers leading the implementation of procedures in their respective spheres of competence of their sub-unit (ibidem). At the lowest levels, instead, two further department are in charge of interacting with partners and suppliers: the Logistics Unit deals with the operational aspects while the Supply Unit is tasked with the commercial dimension, and also with the evaluation of supply chain's security, even when occur an interruption by assessing partners' business continuity level (Norrman and Jansson, 2004). To conclude, Chief Executive Officer (CEO) oversees the overall performance while the Board of Directors and the Audit and Compliance Committee are jointly responsible for the efficacity of the implemented measures and processes (Ericsson Annual Report, 2019). Ultimately, the establishment of a pyramidal structure allows to decentralize the operational activities as a means to ensure an efficient management of the supply chain's extent and complexity. Indeed, each department and sub-unit individually runs the functioning and the coordination of their tasks to maintain the optimal balance between risks' exposure or vulnerability, and cost-effectiveness linked to preventive and safeguard measures (Norrman and Jansson, 2004). In achieving this result, considerable efforts have been made in shaping a real risk culture, intended as an attitude towards a deeper understanding of threats affecting the decision-making process.

THE RISK MANAGEMENT CYCLE

This section analyses the practical functioning of the Ericsson's risk management cycle has divided into 4 steps. The first phase begins with a bottom-up risks' identification process that has been done by each manager within its own area of responsibility (Ericsson Annual Report, 2019). For instance, it could be used a model that summarized several typologies of threats already known and classified, as a tool able to recognize emerging trends and to ensure all sectors to be considered. During this first step, each potential risk linked to the supply chain is examined separately through different techniques, such as the Fault Tree Analysis (FTA) or the Event Tree Analysis (ETA), namely a graphic representation that allows to consider and visualize how a set of failures could spread throughout the system (Norrman and Jansson, 2004). Lastly, this process is strictly linked to a further top-down risks' identification phase where the Group Risk Management Function refines and approves the initial assessment with other external experts (Ericsson Annual Report, 2019). Thus, the Swedish multinational can realize a complete representation of the challenges to cope with, and to visualize all the elements of its supply chain. By examining the flow of trade, it could be possible to define more precisely either the priority areas on which to intervene as well as the risk's sources, for example a component or a production plant.

After having identified the causes related to potential threats, the goal pursued in the second phase of risk analysis involves reaching an even deeper awareness of the probability of occurrence and the impacts which could result (Norrman and Jansson, 2004). Firstly, the reliability and security of partners are assessed in terms of quality and availability of suppliers. Afterwards, the predictable negative effects linked to each threat are estimated considering different dimensions: financial, reputational, commercial, as well as external environment and internal vulnerabilities. Ericsson has developed a set of measures that help analysing strategic uncertainties, including a calculation of the consequences based on the time range required to restore the system, expressed through the Business Recovery Time (ibidem). The latter is multiplied by the gross profit in order to obtain the corresponding financial impact of the risk. Thus, the Business Interruption Value is quantified, and the firm has a more comprehensible measure when priorities are to be assigned and solutions are to be taken (ibidem). Usually, the results of this step are reported on a heat map which allows to graphically illustrate the impact values and the probabilities related to each identified threat.

The third phase is the risk treatment: in this phase hypothetical actions to mitigate negative effects are discussed (Ericsson Annual Report, 2019). This is the central moment of the whole process, at the end of which the optimal strategy is selected. The decision arises based on the work carried out by each manager within its department or sub-unit. In fact, they are required to draw up standard models describing risks' profiles and several preventive or mitigating measures of which their costs and effectiveness are specified (Norrman and Jansson, 2004). The different treatment options suggested from the lowest levels are, then, reworked by the Group Risk Management Function to identify those that are more effective. Usually, the solution falls into one of the following types emerged from the literature review (Chopra e Sodhi, 2004; Ravindran e Warsing, 2013): (i) to take the risk, for instance in the case of low probability and low impact events or when the proposed solution is considered too burdensome compared to the problem; (ii) to transfer the risk, making it fall on the partners or by concluding an insurance contract which covers potential damages; (iii) to avoid the risk, meaning reduce the occurrence probability related to negative events by implementing preventive measures aim at facing the existent vulnerabilities or mitigate the severity of the effects linked to threats, for instance, by means of Early Warning System able to timely inform the emergence of new source of instability; (iv) to remove the risk, by redesigning industrial structures, productive procedures or corporate strategies choosing alternatives that guarantee new exposure value to threats or to damages and losses. Finally, it should be noted that the final decision will be the results of a cost-benefit analysis in relation to each treatment option submitted (Chopra and Sodhi, 2004).

Following the strategy's selection, its effectiveness is continuously evaluated and, if necessary, corrective solutions need to be implemented. During this monitoring and final phase, the Group Risk Management Function provides management actions that meet the evaluation requirements contained in international standards and coordinates controlling activities between each manager at the lower level (Ericsson Annual Report, 2019). An analysis to which the various partners of the supply chain are also subjected, in particular their ability to fulfill the commitments undertaken in terms of corporate security (Norrman and Jansson, 2004). In this respect, Ericsson requires its suppliers to develop business continuity plans, to identify alternative resources or production plants, and to report any kind of incidents or other negative circumstances. According to this principle, the Swedish company has also set up specific programs to cope with supply chain's interruption cases aiming at increasing business operational continuity (ibidem). Essentially, these are guidelines describing how to behave in certain situations, what safeguards mechanism need to be activated, and who are responsible for crisis management. Thus, these measures are associated with the risk management cycle and they could be considered as additional tools to protect company's assets and to react to external uncertainties. To conclude, this outlined approach takes on a proactive value to the extent that this cyclicity is maintained through updating meetings and reports.

DISCUSSION

Based on the analysis carried out, it has been possible to describe in detail how a risk management process actually works. The Ericsson's method allows to improve supply chain's agility and resilience developing a proactive attitude at the organizational level. The result that emerges is the capacity to mitigate the negative effects linked to risks, to which it relates the efficient management of incidents or interruptions. Moreover, it should be noted that investments to achieve a strengthening of the supply chain should not be perceived as mere costs since, working to prevent harmful events, firms will suffer fewer losses both in terms of profits and reputation.

In the attempt to address the whole spectrum of relevant elements, this model is configured as holistic approach that goes beyond the internal dimension by regulating, at the same time, partners and suppliers' interaction modalities. Thereby, its purpose is to face risks from different perspectives bolstering the collaboration within the supply chain, for instance, achieving ISO certifications which guarantee a constant control over company's quality and security. Furthermore, it shows an increasing attention to the problem of the interconnection of threats and the importance of developing concerted initiatives to address them both at national and international level. This is exactly where the Ericsson Response program fits along with other numerous cooperation agreements between private companies and humanitarian organizations in an effort to increase communities' resilience by providing solutions to pre-existing vulnerabilities that have the potential to generate harmful phenomena. Indeed, hazards become risks when a deep understanding of vulnerabilities and resources' exposure is reached, aimed at mitigating those weaknesses. In essence, global supply chain's security lies in its complexity which in turn requires the ability to intervene simultaneously in different areas. Sustainability and corporate social responsibility's principles could be integrated into the corporate strategy for the purpose of understand and proactively manage threats, and also the resulting environmental, social, and economic impacts. In addition, such awareness provides the ability to seize new opportunities leveraging reliable and constantly updated data when drawing up companies' orientations.

RISK'S INTERACTION AND CORPORATE RESPONSIBILITY

The following section explores the issue of malfunctioning of international logistics, meaning at introducing a new element in the overall picture of risk management associated to natural disasters. For instance, the bond between private and private sector will be explored, focusing on the awareness of the negative impact that would affect both parties involved in case of tragic, unpredictable events. As Grimm (2013) claims, the companies' security is directly related to the society's subsistence. The recognition of the interactive nature of the threats in a modern dynamic context implies the companies' urgence to undertake, through solidarity initiatives, the implementation, on one side, of the population resilience, while, on the other, the strengthen of the business continuity's capacity (McKnight and Linnenluecke, 2016).

THE RIPPLE EFFECT OF GLOBAL RISKS

The financial-economic globalization, together with the constant increase of cash and goods' flow, leads a set of inevitable negative aspects, among which the possibility to trigger a ripple effect as soon as it occurs whether an adverse event or a failure in the operative mechanism of the supply chain (Cristopher and Lee, 2004). In the international supply chain, whatsoever accident that cause a temporary interruption of commercial flows, could finally result in great economic losses and sequential damages, due to the consequent occur of several secondary events that would amplify the impact of the former accident. Thus, it is not worth considering the factors of risk as isolated: these factors must be analyzed taking into consideration the network of interdependency to which they belong (Kwak et al., 2018). Furthermore, this correlation is perfectly in line with the characteristic of the global supply chains, where complexity is at the core of the relations between the several parts involved.

In order to investigate the foremost causes behind the paralysis of the supply chain, Kwak et al. (2018) refer that the threats' effects are increased by the interaction of different sources of risk, identifying in this respect four levels of vulnerability in the international logistic: (i) the flow of value, meaning that whatsoever concern the production of goods and services is influenced by the perception of threats intended as variation in time, quality or costs; (ii) the reliability of relations among partners depends instead on the dimension of the system and its capacity to elaborate and communicate critical information to the overall functioning; (iii) the activity and infrastructure securing of the supply chain refers to the risks deriving from the planification and management of the supply network; (iv) the external environment refers, to conclude, on the threats associated to the occur of natural disaster, but also to the distress consequence of human impact, generating a negative impact in the movement of goods (Hausman et al., 2005). It is important to underline how this vulnerability refers to phenomena which are usually considered of low frequency but of high impact (Tang, 2006) since, interacting with high risks, they produce repercussions all over the logistic infrastructure. According to this analysis, it is necessary to intervene through the application of reactive and proactive measure, aim at minimizing or pre-empting the possibility of an initial event that would produce far reaching consequences. In particular, Kwak et al. (2018) consider feasible the reduction of the ripple effect through an intervention on the deficiency correlated to the information and control systems all over the supply chain. In case of a damaging event or an interruption, the reliability of the suppliers is crucial to correctly handle the sudden and to adapt the commercial infrastructure. The explanation lies in the fact that companies must rely the more and more in the capacity of data processing of further actors, also operating in the chain (Ibidem). The developing of important relations with commercial partners and Logistic Service Providers (LSP) is configured as a mitigation strategy, as far as the risk management system incorporates a correct burden-sharing of information and efficient relations between the decision-makers, avoiding that from a single phenomenon it could arise misunderstandings or communications distortions, causing a set of untoward situations (Bode et al., 2011). Among the means to reach this result, it is possible to count the international standards, as the ISO norms mentioned above: they assure the respect of current legislations furnishing useful elements to better implement adequate control systems. In other words, the ability to visualize the chain single elements together with effective collaborative relationship between the various sectors represents necessary requirements to face those risks generated in the logistic international network, crucial since they could compromise the overall performances.

SOCIAL RESPONSIBILITY OF THE PRIVATE SECTOR

It is possible to claim that a better understanding of the root causes of potential damaging events, could be obtained through the identification and recognition of the dynamic nature of the risk, so that those in charge would be able to assign a correct value in order to elaborate an accurate strategy. In fact, several measures could be implemented in order to increase the flexibility, the agility and adaptability of the supply chain, since they represent the essential element to make the supply chain resistant and resilient, reducing the time of response and re-elaboration of the critical function at its minimum (Kwak et al., 2018). However, the phenomenon of the threats' interconnection presents a further implication since it involves the need to develop a wider approach, in order to solve not only the internal vulnerabilities but also and foremost the long-term ones to face the most far-reaching factors of instability. From this perspective, the multinationals are clearly interested in protecting their goods and resources from the increasing risks' exposition in several world regions, where they operate. The principles of business continuity, which provide the correct means to strengthen the organization's infrastructure, could be extended in order to establish a common substructure to develop a new effort for the risk-management, e.g. intervening to reduce damages in the critical infrastructures of a specific country (Herbane et al., 2010).

The effectiveness of this methodology in the business perspective has been highlighted with the Ericsson's case. Instead, the following section explores the companies' network of information-sharing, which result essential to the elaboration of assistance programs for the response and the preparation in view of sudden natural disaster. It is important to specify that the private sector is more inclined toward economic advantages, even when it commits itself helping its clients to recover from an adverse event, conceiving solution to reduce the repercussion and preserving the business competitiveness. Nevertheless, McKinght and Linnenluecke (2016) claimed that in time of crisis the communities depend on the resilience of those in charge of the management of the essential infrastructure of the country. These companies are able to stabilize the crisis situation, providing services and further materials to vulnerable people, saving their employees' occupation and gaining profits supporting also in the reconstruction stage (Ibidem). In fact, solid infrastructural systems are essential to minimize the impact of extreme events. Nevertheless, this is exclusively one aspect, since the private sector would play a further role: for instance, the companies could benefit from their experience and presence in certain territories to link the response activities to the concrete needs of the community (Ibidem). They build up reliable relations between the parties involved through information-sharing and resources or the elaboration of ideas to face and solve the difficulties encountered by the populations affected, through medium-long term partnerships with governmental or humanitarian organizations (Maak, 2007). With respect to the second case study, these cooperation accords play an enhanced role, elaborating rescue programs to face the most complex emergencies. Briefly, it can be considered as a mean to reach a double objective: on one side, operating in instable areas in order to avoid the main damaging effects on the global supply chain of a disaster; on the other side, increasing the competences and resources available to the local communities in order to reduce their exposition and vulnerability to following extreme phenomena.

RESPONSE'S COLLABORATIVE PROGRAMMS TO NATURAL DISASTERS

Although governments are primarily responsible during an emergency, Chen et al. (2013) have suggested that they do not have the necessary means and expertise to cope with such crisis situations. Instead, since private sector holds a large number of critical infrastructures within countries, policymakers have the task of involving companies to jointly develop suitable solutions (ibidem). The literature review has revealed three types of commitment by enterprises in the disaster response phase: (i) emergency engagement actions through business continuity management's practices (Herbane et al., 2010). (ii) Philanthropic initiatives, based on Corporate Social Responsibility's principles, and characterized by in-cash or in-kind donations; in particular, they allow to obtain benefits in financial terms, and above all of reputation (Tilcsik and Marquis, 2013). (iii) Private-public collaborations during the whole disaster management cycle where firms are involved in the development of specific and innovative solutions to facilitate relief operations (Shepherd and Williams, 2014). Usually, enterprises' offered services consist in the recovery of sensitive data, in the rehabilitation of facilities or emergency systems as well as several form of consultancy. Research in this direction shown the advantages for companies which have the possibility to exploit new opportunities and broaden their commercial range while protecting its own assets, for instance by providing infrastructure rehabilitation services to develop new relationships with stakeholders or to strengthen the existing ones (McKnight and Linnenluecke, 2016). In the view of an increase in frequency and intensity of adverse natural phenomena, therefore, private sector shows a tendency to greater involvement in socially oriented programs to face global vulnerabilities. These latter often take the form of partnership with international agencies dealing with emergency management such as that carried out by Ericsson with the Emergency Telecommunications Cluster (ETC), namely a worldwide network of organizations working together to provide shared communication services during the crises in which the World Food Programme (WFP) acts as UN Logistic Cluster leader.

Established in 2000, the Ericsson Response is a relief program based on providing and restoring telecommunication services in the aftermath of a natural disaster. The main task is to install and maintain Internet connection until local infrastructures are reactivated, allowing humanitarian operators to leverage several technologies and devices to reach and communicate with the population in affected areas (Ericsson Fact Sheet, 2020). On the occasion of the 20th anniversary of this initiative, the Swedish company has revealed some data about the mission's effectiveness. It is entirely based on volunteer employees including engineers, communication experts, and other kind of specialists: starting from the operations in Algeria affected by heavy flooding in 2000, the team has intervened in over sixty humanitarian crises in forty countries, such as Tajikistan (severe droughts in 2000), Haiti (the 2010 earthquake, the 2011 cholera outbreak, and the 2016 hurricane), Puerto Rico and Dominican Republic (the 2017 hurricanes), and Philippines (the 2013 typhoon). These examples demonstrate the program's success as confirmed by the most recent humanitarian action of 2019 when the rescue team was engaged in Bahamas in the aftermath of the Dorian hurricane (ibidem).

Broadening this analysis perspective, aim at verifying the effectiveness of humanitarian international partnerships, the final section describes a second case study based on the Logistics Emergency Teams (LETs), exploring the benefits that could arise from cross-sector cooperation in terms of operational performances and safety advantages to local communities.

EMERGENCY MANAGEMENT'S CROSS-SECTOR COLLABORATION: THE LOGISTICS EMERGENCY TEAMS CASE

This section investigates humanitarian collaboration by discussing of one of the largest international programs, namely the Logistics Emergency Teams (LETs) established within the structure of UN Logistics Cluster headed by the World Food Programme (WFP). This system guarantees speed, agility, and a greater operational efficiency to address global crisis scenarios by providing solutions to the humanitarian supply chain thanks to its several branches. From the experiences accumulated over the years and given the increase in the frequency and intensity of natural disasters, the need to involve the private sectors has been warned through partnerships that allow the integration of enterprises in the humanitarian community in pursuit fixed objectives. Due to supply chain and business continuity management's expertise, it is no coincidence that logistics firms were the first to join new forms of socially oriented cross-sectorial cooperation. The following analysis illustrates the main features of the program, the successful elements but also the critical issue emerged from this agreement by means of institutional sites, reports, official documents as well as specific academic research.

The need to integrate business skills within the emergency management international system was satisfied in 2005 when took shape the LETs initiative, under the auspices of World Economic Forum. It was the result of combined efforts of some multinational companies in the logistics sector who recognized the crucial role of this kind of services during a crisis and began to study ways to increase their efficacy as well as the coordination between different institutional levels (Stadler and Van Wassenhove, 2012). On one side, these enterprises had those resources and skills that could bring significant additional value to relief operations. On the other side, the need to articulate a collaboration typology that does not involve an increase in the complexity of operations, characterized by the large number of actors usually involved, was warned. In this perspective, emergency logistics teams were set up under the co-supervision of the companies themselves and the WFP but integrated in the UN Logistics Cluster's structure (ibidem). Currently, the LETs are made up of four enterprises among the largest in the sector, namely United Parcel Service (UPS), A.P. Moller-Maersk, Agility, and DP World, whose main objective is to provide support to humanitarian operations following the occurrence of a large-scale natural disaster (LET, 2019).

The partnership is deployed through several dimensions going from the preparation and mitigation of potential adverse effects to anticipate an extreme event until the direct participation in emergency response operations by providing specialized staff and a range of goods and in-kind services. Moreover, three clearly defined activation criteria were established: (i) companies contribute through their core competencies on *pro-bono* based, (ii) after receiving a formal request from the WFP as Logistics Cluster leader, and (iii) only in the case of a natural disaster involving more than five hundred thousand people (Stadler and Van Wassenhove, 2012). However, given the evolution in the global humanitarian scenario in terms of operations' magnitude and duration, starting from 2017, LETs' members have decided to expand their mission: from response activities to sudden-onset natural disasters, such as earthquakes or hurricanes, even to recognize the need to also intervene during slow-onset complex emergency, for instance health crises or plagues such as drought or famine, and even through the development of prevention or mitigation programs of various forms of instability in risky areas (LET, 2019). Again, this evolution demonstrates the private sector's desire to assume an increasingly decisive role in developing strategies and goals pursued by the international humanitarian community. Demonstrating the availability to engage in a long-term perspective towards government and local communities through this initiative, LETs' members can therefore appear and be perceived as reliable and responsible actors, in turn, benefiting from a positive corporate image and other commercial benefits. In these fifteen years, the LETs have intervened in 22 natural disasters and crises around the world among which it is possible to count the response activities to the 2008 cyclone Nargis in Myanmar, the 2010 severe flooding in Pakistan, the 2016 hurricane Matthew in Haiti as well as other operations to face deeper instability as the causes of the heavy famine affected the Horn of Africa area in 2011 or the assistance to refugees in Bangladesh, Yemen, Syria and Iraq in 2018 (LET, 2019).

THE RISK-MANAGEMENT OF THE TRANSPORT SYSTEMS

Enforcing the risk-management approach explored in the first study case, to the uncertain and dynamic context of the rescue agencies' operations, it is possible to demonstrate the operational efficacity arising from a partnership initiative, as that of LETs. The logistic dimension results a crucial asset to the development of an efficient supply chain, implying the necessity to reach a deep understanding of those elements that could compromise the functioning of the overall infrastructure. It follows that the management of the transport operations and of the whole organizational and infrastructural risks, connected thereto, represents a central challenge to guarantee the success of the relief efforts. Following this path, a recent research on the earthquake that has hit Nepal in 2015, identified a set of challenged and obstacles coped by the rescue supply chain, characterized by coordination and information sharing issues (Baharmand et al., 2017). In particular, interviewing several operators in the field, it emerged a classification related to transport risk within the country itself (Ibidem): (i) delay in delivery, e.g. caused by adverse weather conditions, traffic density or infrastructures damaging; (ii) loss of rescue material, due to accidents or aftershocks; (iii) insufficient cargo capacity or unavailability of the adequate means of transports; (iv) deterioration or damaging of goods, for instance in case of equipment requiring specific methods of preservation, as medicament or food; (v) market volatility, meaning an increase of prices of certain either goods or services in crisis situations; (vi) unreliable information, as the lack of sources or the inactivity of adequate technologies; (vii) unethical behaviors, due to the absence of trained staff to face emergency contexts.

The positive influence deriving from the intersectoral collaboration between humanitarian organizations and specialized companies, fits in this management dimension. The enterprises could play a crucial role in facing the eventual malfunctioning of the infrastructures, placing at the disposable either their competences and experiences in the field of trade, or their variety of facilities among which means of transport or vanguard tracking technologies. The main value of this cooperation is perceived as related to those risks previously identified whereas the ability to manage them influenced the overall trend of the intervention. For instance, Baharmand et al. (2017) demonstrated the advantages reached in the field, among which it is important to outline the ability to improve the strategic planning thanks to an efficient allocation of resources, to strengthen the cargo and drivers' security, and to take advantage of the presence in the territory and the trade network established in order to guarantee either a reliable supply of material or the correct flow of information. Consequently, the exploitation of logistic services' societies represents a mitigation strategy within the disaster management cycle, as far as it enables to implement the efficiency, flexibility and reactivity of the supply emergency chain's performances (ibidem). Moreover, it demonstrates the importance of the predisposition of a set of cross-sector accords and agreements to recourse in case of crisis. A requirement satisfied by the LETs' initiatives and by the involvement of the private sector, in order to stimulate a process of reciprocal learning and provide significant contributes at the operational level. In particular, filling the gaps of humanitarian organizations whereas necessary, these partners could make use of their risk-management and business continuity knowledges in order to face the several obstacles and continuous challenges imposed by an extremely challenging external environment.

DISCUSSION

In accordance with this analysis, it is clear how the existent literature and the operational reality underline the relevance of these partnerships among organizations that share similar characteristics, although in different scope. The recognition of the rising interconnection of instability factors has provided the impetus to initiate a phase of dialogue between various actors at the national and international level: renewed awareness accrued also by private enterprises that decide to engage in solidarity projects, finding themselves in the ideal conditions to expand their own strategic objectives. It results in the developing of proactive orientation, aim at intervening in a decisive manner and in the long-term in favor of the local community, aware that the social-economic well-being comes along with an improved security. The initiative of the Logistics Emergency Teams perfectly fits in this direction and provide a correct response to the need to strengthen the society and those areas affected by the extreme events. This occurs in the light of the exchange of knowledges and resources between organizations during the whole emergency management cycle. Through the application of the continuity business practices in the humanitarian operations, it has been valued the way in which these implementations could contribute to face the threats derived from an uncertain and dynamic environment. Certainly, the cross-sector cooperation guarantees remarkable benefits, in particular in terms of time, costs and capacity to reach an increased amount of people in distress (Nurmala et al., 2017). Nevertheless, the risk management as methodology able to inform and lead the decision-making process of each either public or private organization, appear to be an additional value, faced with the tendency of an increase in the intensity and frequency of disasters, which reflect deeper vulnerabilities, although their sudden natural origin.

CONCLUSION

The purpose of this study is to investigate risk models and strategies for the security management of global supply chains. In addition, the research aims at highlighting the common elements between public and private organizations, meaning the way they address crisis scenario, with a specific focus on the issue of interaction between natural phenomena and preexisting vulnerabilities. The results achieved have provided several insights on a topic of recent interest although some challenges have been raised which need to be considered.

Albeit in different areas, humanitarian organizations and enterprises shown several continuity elements that have been investigate in relation to the logistics dimension, above all due to its crucial role during relief operations. Moreover, the latter allows to point out opportunities for mutual learning that may arise from cross-sector collaboration in terms of sharing resources and knowledge. The theoretical and practical evolution of the concepts of supply chain management has placed the bases to deepen managerial and operational aspects, subsequently applied to emergency management activities in order to optimize the use of resources and improve their overall efficiency. In particular, the extension of these commercial principles to humanitarian supply chains has shown important advantages in terms of agility, reactivity, and reliability of the system with which they respond to a crisis situation.

The analysis of the disaster management cycle's characteristics has demonstrated how the trend towards the application of a wider proactive approach, aim at pre-empting the emergence of new instabilities sources through the elaboration of structural mitigation measures, has arisen. The humanitarian organizations have developed a fundamental attitude to the strategic planning as an instrument to inform the decision-making process about the factors of risk. This reaction capacity to external stimulation has been achieved thanks to the adoption of principles and practices aim at increasing the internal organizational flexibility. This dimension is gaining importance also in the private sector, considering the threats inherent in operating on a global scale and the role played by the logistic in generating profits in this kind of conditions. Indeed, the precondition at the core of the transboundary trade is that goods can be moved in a safe and reliable manner, reaching their destination at an overall cost lower than the one of equivalent goods locally produced. Therefore, the management of threats, not only in order to protect but also to improve the supply chain performances, represents a fundamental challenge to which several enterprises have been able to response with adequate solution, resorting to the business continuity managements concepts.

As can be seen from the analysis conducted in relation to the model adopted by Ericsson, an effective risk management system can be set up using recommendations and guidelines of international regulatory standard such as the ISO discipline on

Enterprise Risk Management. In particular, an important aspect has emerged where the ability to intervene on the information flow and on the relational modalities between different partners ensure a constant monitoring of the whole supply chain. In combination with security measures and contiguity plan, risk management practices could make a significant contribution to the ability to reduce or mitigate the adverse effects of global supply chain paralysis.

In an effort to deal with the problem of the threats' interconnection, the involvement in various form of partnership with humanitarian agencies shows how companies could play a crucial role in emergency management where philanthropic and commercial intentions overlap each other through the Corporate Social Responsibility. Presenting the case of LETs partnership, the current international orientation clearly emerges where the objective of increasing the resiliency of communities most exposed to adverse events is linked to emergency response actions. This cross-sector cooperation provides several insights on sharing resources, demonstrating the effectiveness of the application of management strategies from private sector to humanitarian operations by ensuring the correct flow of information and appropriate relations between decision-makers. Finally, literature and operational reality highlight the validity of collaboration aimed at the exchange of best practices in achieving common goals.

The research perspective taken by this study allowed to explore a theme of recent interest. The results achieved underline the usefulness of an analytical process that interpret and integrates relevant information in a logical combination able to outline a spectrum of viable alternatives in relation to a specific problem. In this regard, such managerial strategies fit into the domain of intelligence analysis by providing information on major threats to structure measures and action plans which minimize risks. A form of knowledge verified and then transformed into a valuable assessment to the extent that it becomes an available operational tool to decision-makers, both in international-humanitarian and private sectors.

However, some limits must be taken into consideration. Firstly, the need to deepen the subject with further empirical research to better identify the opportunity to share experiences for value creation in both sectors. Moreover, it could turn out to be useful the further investigation of impact on global supply chains of other types of adverse events such as those of human origin, for instance, terroristic attacks as well as social and political instabilities. Nevertheless, risk management process as a methodology capable of informing and guiding decision-making stages of any public and private organization seems to be a crucial added value in the face of the constantly increasing trend of natural disasters which reflect deeper vulnerabilities, although originating from natural hazards.

REFERENCES

Annual Disaster Statistic Review from the Emergency Events Database EM-DAT. (2011). *Centre for Research on the Epidemiology of Disasters (CRED)*. Available at https://www.emdat.be/publications

Baharmand, H., Comes, T., & Lauras, M. (2017). Managing in-country transportation risks in humanitarian supply chains by logistics service providers: Insights from the 2015 Nepal earthquake. *International Journal of Disaster Risk Reduction, 24*, 549–559. doi:10.1016/j.ijdrr.2017.07.007

Birkie, S. E., & Trucco, P. (2020). Do not expect others do what you should! Supply chain complexity and mitigation of the ripple effect of disruptions. *International Journal of Logistics Management, 36*(7), 27–29.

Bode, C., Wagner, S. M., Petersen, K. J., & Ellram, L. M. (2011). Understanding responses to supply chain disruptions: Insights from information processing and resource dependence perspectives. *Academy of Management Journal, 54*(4), 833–856. doi:10.5465/amj.2011.64870145

Chen, J., Chen, T. H. Y., Vertinsky, I., Yumagulova, L., & Park, C. (2013). Public-private partnerships for the development of disaster resilient communities. *Journal of Contingencies and Crisis Management, 21*(3), 130–143. doi:10.1111/1468-5973.12021

Chopra, S., & Sodhi, M.S. (2004). Managing risk to avoid supply chain breakdown. *Massachusetts Institute of Technology Sloan Management Review, 46*(1), 53-61.

Christopher, M., & Lee, H. (2004). Mitigating supply chain risk through improved confidence. *International Journal of Physical Distribution & Logistics Management, 34*(5), 388–396. doi:10.1108/09600030410545436

Cozzolino, A. (2012). *Humanitarian logistics. Cross-sector cooperation in disaster relief management*. Springer Briefs in Business Series. Springer.

Cozzolino, A. (2013). *L'agilità nella logistica delle emergenze. Come le imprese apprendono la gestione dell'imprevedibile dalle organizzazioni umanitarie nelle operazioni di primo soccorso.* Convegno Aidea, Lecce, "Il ruolo dell'azienda nell'economia. Esiste un modello aziendale orientato alla crescita?

Cozzolino A. (2014). Agilità nella logistica delle emergenze. Le imprese apprendono dalle organizzazioni umanitarie. *Sinergie, Italian Journal of Management, 95*, 75-98.

Cozzolino, A., Rossi, S., & Conforti, A. (2012). Agile and lean principles in the humanitarian supply chain. The case of the United Nations World Food Programme. *Journal of Humanitarian Logistics and Supply Chain Management, 2*(1), 16-33.

Ericsson Response Fact Sheet. (2020). *Breaking the energy curve*. Author.

Grimm, D. (2013). Whole community planning: Building resiliency at the local level. *Journal of Business Continuity & Emergency Planning, 7*, 403–438. PMID:24578027

Harvard Business Review. (2001). *Risk Management in a Time of Global Uncertainty*. Author.

Hauser, L.M. (2003). Risk-adjusted supply chain management. *Supply Chain Management Review, 7*(6), 64-71.

Hausman, W. H., Hau, L., & Subramanian, U. (2005). *Global logistics indicators, supply chain metrics and bilateral trade partners.* World Bank Policy Research Working Paper No. 3773. doi:10.2139srn.869999

Herbane, B., Elliott, D., & Swartz, E. M. (2010). Business continuity management, second edition: A crisis management approach. New York, NY: Routledge.

Ilha A. (2011). The humanitarian relief chain. *South East European Journal of Economics & Business, 6*(2), 45-54.

International Organization for Standardization (ISO). (2009). *Risk management – principles and guidelines.* ISO 31000:2009. Available at https://www.iso.org/obp/ui/#iso:std:iso:iwa:31:ed-1:v1:en

International Organization for Standardization (ISO). (2012). *Societal security – business continuity management systems – requirements.* ISO 22301:2012. Available at https://www.iso.org/obp/ui/#iso:std:iso:22301:ed-1:v2:en

Ishida, S., Montshiwa, A. L., & Nagahira, A. (2016). Modifying business continuity plan (BCP) towards an effective auto-mobile business continuity management (BCM): A quantitative approach. *Journal of Disaster Research, 11*(4), 691–698. doi:10.20965/jdr.2016.p0691

Kanter, R. M. (1999). From spare change to real change. The social sector as beta site for business innovation. *Harvard Business Review, 77*(3), 122–132. PMID:10387574

Kleindorfer, P. R., & Saad, G. H. (2005). Managing disruption risks in supply chains. *Production and Operations Management, 14*(1), 53–68. doi:10.1111/j.1937-5956.2005.tb00009.x

Kwak, D. W., Rodrigues, V. S., Mason, R., Pettit, S., & Beresford, A. (2018). Risk interaction identification in international supply chain logistics: Developing a holistic model. *International Journal of Operations & Production Management, 38*(2), 372–389. doi:10.1108/IJOPM-03-2016-0121

Lee, H. (2004). The triple-A supply chain. Harvard Business Review on Supply Chain Management, 82(10), 102-112.

Logistics Emergency Team (LET). (2019). *Annual Report.* Available at https://logcluster.org/document/let-annual-report-2019

Long, D., & Wood, D. (1995). The logistics of famine relief. *Journal of Business Logistics, 16*(1), 213–229.

Maak, T. (2007). Responsible leadership, stakeholder engagement, and the emergence of social capital. *Journal of Business Ethics, 74*(4), 329–343. doi:10.100710551-007-9510-5

McKight, B., & Linnenluecke, M.K. (2016). How firm responses to natural disasters strengthen community resilience: A stakeholder-based perspective. *Organization & Environment, 29*(3), 290-307.

Norrman, A., & Jansson, U. (2004). Ericsson's proactive supply chain risk management approach after a serious sub-supplier accident. *International Journal of Physical Distribution & Logistics Management, 34*(5), 434–456. doi:10.1108/09600030410545463

Nurmala, N., de Leeuw, S., & Dullaert, W. (2017). Humanitarian-business partnerships in managing humanitarian logistics. *International Journal of Supply Chain Management, 22*, 82-94.

Oglesby, R., & Burke, J. (2012). *Platforms for private sector-humanitarian collaboration.* King's College London.

Ravindran, A. R., & Warsing, D. P. J. (2013). *Supply chain engineering: Models and applications.* CRC Press.

Ericsson Annual Report. (2019). *Building a stronger company long term.* Author.

Shepherd, D. A., & Williams, T. A. (2014). Local venturing as compassion organizing in the aftermath of a natural disasters: The role of localness and community in reducing suffering. *Journal of Management Studies, 51*(6), 952–994. doi:10.1111/joms.12084

Stadler, L., & Van Wassenhove, L. N. (2012). The Logistics Emergency Teams: Pioneering a new partnership model. Case Study, INSEAD, European Case Clearing House, ECCH (Ref. No. 712-033-1).

Tang, C. S. (2006). Perspectives in supply chain risk management. *International Journal of Production Economics*, *103*(2), 451–488. doi:10.1016/j.ijpe.2005.12.006

Thomas, A. (2003). *Humanitarian Logistics: Enabling Disaster Response*. White paper, Fritz Institute, San Francisco, CA.

Thomas, A., & Kopczak, L. (2005). *From logistics to supply chain management: the path forward in the humanitarian sector*. Fritz Institute.

Tilcsik, A., & Marquis, C. (2013). Punctuated generosity: How mega-events and natural disasters affect corporate philanthropy in U.S. communities. *Administrative Science Quarterly*, *58*(1), 111–148. doi:10.1177/0001839213475800

Tomasini, R. M., & Van Wassenhove, L. N. (2004). *A framework to unravel, prioritize and coordinate vulnerability and complexity factors affecting a humanitarian response operation*. INSEAD working paper, No. 20 04/41/TM, INSEAD Fontainebleau, France.

Tomasini, R. M., & Van Wassenhove, L. N. (2009). From preparedness to partnerships: Case study research on humanitarian logistics. *International Transactions in Operational Research*, *16*(5), 549–559. doi:10.1111/j.1475-3995.2009.00697.x

Trunick, P.A. (2005). Special report: delivering relief to tsunami victims. *Logistics Today, 46*(2), 1-3.

UNDRR United Nations office for Disaster Risk Reduction. (2019). *Global assessment report on disaster risk reduction.* Available at https://www.undrr.org/publication/undrr-annual-report-2019

United Nations. Glossary. (1992). Internationally agreed glossary of basic terms related to disaster management. U.N. International Decade for Natural Disaster Reduction.

Van Wassenhove, L. N. (2006). Humanitarian aid logistics: Supply chain management in high gear. *The Journal of the Operational Research Society*, *57*(5), 475–489. doi:10.1057/palgrave.jors.2602125

Ye, L., & Abe, M. (2012). The impacts of natural disasters on global supply chains. Working Paper Series No. 115, June 2012 in Asia-Pacific Research and Training Network on Trade (ARTNeT), Bangkok, ESCAP.

ENDNOTE

1 Newspaper article, Repubblica. (2009). Available at https://www.repubblica.it/2009/07/sezioni/cronaca/nuova-influenza/virus-popolazione/virus-popolazione.html

Chapter 4
State of the Art of XR Management Training Applications

Irene Maria Gironacci
Swinburne University of Technology, Australia

ABSTRACT

Recent advancements in extended reality (XR) immersive technologies provide new tools for the development of novel and promising applications for business. Specifically, extended reality training applications are becoming popular in business due to their advantages of low cost, risk-free, data-oriented training. Extended reality training is the digital simulation of lifelike scenarios for training purposes using technologies such as virtual reality, augmented reality, mixed reality. Many applications are already available to train employees to develop specific technical skills, from maintenance to construction. The purpose of this chapter is to review the emerging XR applications developed for management training. Specifically, this chapter will focus on the training of some key skills in management such as leadership, problem solving, emotional intelligence, communication, and team working.

INTRODUCTION

Management training consists of programs and courses to train managers. It focuses on training individuals with management skills such as project management, leadership, emotional intelligence, stakeholder management, communication, team working, problem solving and decision making, strategic thinking, stress management, commercial awareness, organization and delegation. Required skills

DOI: 10.4018/978-1-7998-7126-2.ch004

for project management include, and are not limited to schedule management, risk management, budget management, performance management, quality management, and negotiation. One of the challenges in the development of performance-based methodologies to measure leadership skills is how to generate real-life situations with tools that collect data about management competencies. Extended Reality is a relevant experimental tool in the management field that investigates new objective-based methods for management skills assessment (Alcañiz, et al., 2018), allowing a user to repeat a training multiple times, and collect data about the user during the training. In this chapter, some examples of Extended Reality applications for developing required skills for managers across several fields of interests are explored.

STATE OF THE ART

The development of XR technologies in recent years has resulted in more accessible, immersive, and affordable solutions that can prove to be effective for the development of management skills with practical implications and uses.

Hickman et al (Hickman et al., 2017) investigated how Virtual Reality could be used to fill the intercultural leadership's gap generated by one's intercultural knowledge, skills, and attitudes in STEM, through the use of experimental learning and pedagogical foundations principles. This project provides an innovative approach for leaders to become global business leaders in STEM fields.

Aseel et al. (Aseel et al., 2019), investigated the benefits of Augmented Reality for optimizing product delivery in the construction sector, demonstrating that Augmented Reality training facilitated the delivery of construction and integration of project teams.

MBA students at the University of St. Thomas, Opus College of Business can develop their ethical leadership skills through a VR simulation (McLeod, 2018).

NEOMA Business School launched a program for students to use Virtual Reality to develop problem solving skills in a challenging business environment (McLeod, 2018).

Conflict management skills can be developed in risk-free realistic immersive situations. In 2019, University of Newcastle investigated the use of VR to train nursing students in conflict management simulations: the application replicates a real world emergency room and asks students to respond to a high pressure mock scenario. Students' heart rates data were also collected during the task, to teach them how to be calm and diffuse high pressure situations (Newcastle University Official Website, 2019).

In 2012, Paul G. Putman (Putman et al., 2012), developed a Virtual Reality simulation for learning styles and conflict management tactics. This simulator explored the relationship between conflict management and personality type, and what type of adult learner might perform best within a virtual leadership simulation.

Management techniques of various types of conflicts – from high to low intensity – and the development of a positive, solution-oriented attitude was investigated by the company 5Discovery (5discovery website).

Figure 1. Conflict Resolution VR program at University of Newcastle
(Source https://www.newcastleherald.com.au/story/6509037/angry-stan-steps-up-to-simulate-the-stress-of-the-emergency-room-for-student-nurses/)

Figure 2. Discovery Conflict Management in VR
(Source https://www.5discovery.com/en/conflict-management/)

Virtual Reality can also be used to help entrepreneurs to give more charismatic investor pitches: Niebuhr and Tegtmeier (Niebuhr and Tegtmeier, 2019) examined a combination of speech-production and perception experiments and determined if and to what extend the acoustic parameters of a charismatic tone of voice can be improved. Their work provided empirical evidence that the traditional rehearsal setting degrades the charismatic tone of voice of a speaker with each new repetition of the investor pitch.

Virtual Speech developed a VR application to increase communication skills and confidence during public speaking, sales pitching, leadership (see Figure 3). Trainee can learn how to: apply leadership and communication techniques to drive business success and become a more effective leader; deliver powerful messages that lead to action and that control emotions when facing resistance; communicate confidently in any environment, from elevator pitches to high stakes presentations that earn buy-in and cooperation.

How do we teach and encourage inclusive behaviours and prepare teams and managers for tough conversations with employees or colleagues about these topics? Strivr created a Virtual Reality application to help managers and employees tackle these questions.

Figure 3. VR Communication Training by VirtualSpeech
(Source https://virtualspeech.com/)

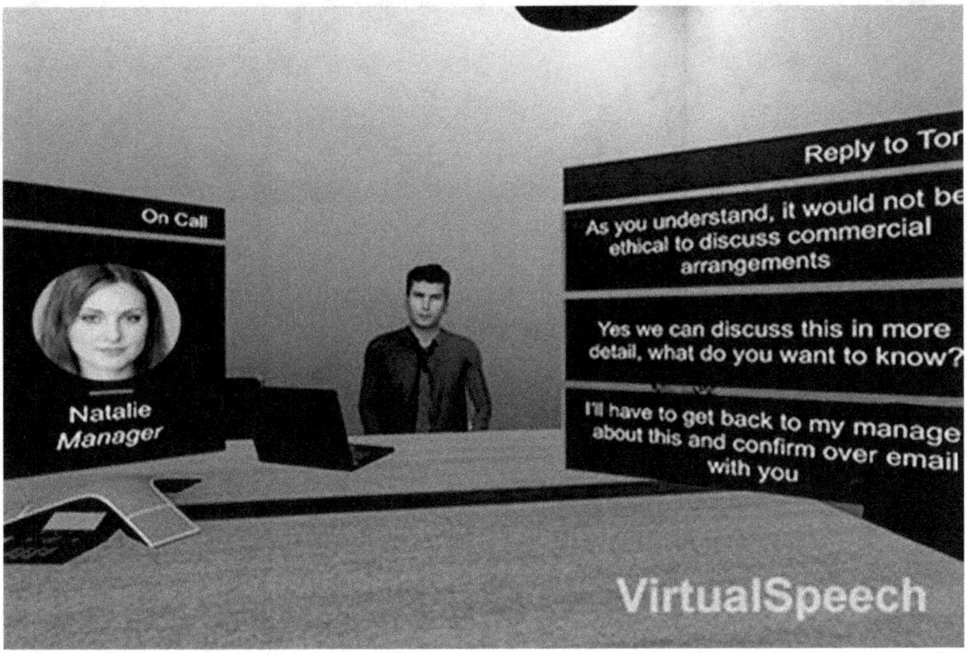

Due to the new reality of working from home, an increasing number of Extended Reality collaboration platform has been used by teams all over the world. Most popular examples of collaborations platform include: Glue (Glue website), Spatial (Spatial.io website), Nvidia Holodeck (Nvidia website). Glue, a VR collaboration platform, is focused on business professionals who need global remote access to a shared team space for efficient collaboration. Spatial, an augmented reality collaboration platform, lets users collaborate, search, brainstorm and share content as if they were in the same room.

Figure 4. Spatial.io
(Source https://techcrunch.com/2020/01/30/spatial-raises-14m-more-for-a-holographic-3d-workspace-app-a-vr-ar-version-of-zoom-or-hangouts/)

A powerful skill for managers is the ability to understand other perspectives, especially when it comes to diversity and inclusion. Empathy can be defined as feeling the same emotion as another observed individual without mixing it with one's own direct experience (de Vignemont and Singer, 2006; Decety and Meyer, 2008; Singer and Lamm, 2009; Decety, 2010).

Extended Reality allows simulations of different scenarios, real-life conditions, environments, and let the user experience different perspectives as well. A study at Standford's Virtual Human Interaction Lab finds that Virtual Reality can help make people more compassionate (Asher et al., 2018). According to the study, people

Figure 5. Developing empathy in VR, Standford's Virtual Human Interaction Lab
(Source https://news.stanford.edu/2018/10/17/virtual-reality-can-help-make-people-empathetic/)

who saw in VR what it would be like to lose their jobs and homes developed longer-lasting compassion towards the homeless.

Hilton uses VR to train their team members to develop empathy towards customers, and handle challenging interactions with customers (Oculus Business, 2020): team members take on the role of guests (a different perspective) in virtual scenarios that include interactions at a front desk, meetings, room service, breakfast, and departure.

CONCLUSION

This chapter presented how Extended Reality technologies can be used to develop management skills, providing a list of use cases both in industry and academia, and highlighted both the advantages offered by these technologies, and the rapid growth of this technology in recent years to provide effective training. In particular, this chapter highlighted how Virtual Reality is used to help users to give more charismatic investor pitches, to collaborate, to develop empathy towards customers, for conflict resolution, to develop problem solving, and to develop leadership skills. Despite many applications are now possible, these technologies comes with some limitations – these includes limited field of view for Augmented Reality and Mixed Reality headsets, episodes of motion sickness in some individuals during Virtual Reality experiences, and the compatibility with eyeglasses only on some specific headsets. As new technologies emerge from the market, it is expected that emerging VR technologies and tools will improve, overcome limitations and further facilitate the training of new emerging managers in the future.

REFERENCES

Alcañiz, M., Parra, E., & Chicchi Giglioli, I. A. (2018). Virtual Reality as an Emerging Methodology for Leadership Assessment and Training. *Frontiers in Psychology*, *9*, 1658. doi:10.3389/fpsyg.2018.01658 PMID:30258378

Asher, T., Ogle, E., Bailenson, J. N., & Herrera, F. (2018). *Becoming homeless: a human experience. ACM SIGGRAPH 2018 Virtual, Augmented, and Mixed Reality*. doi:10.1145/3226552.3226576

De Vignemont, F., & Singer, T. (2006). The empathic brain: How, when and why? *Trends in Cognitive Sciences*, *10*(10), 435–441. doi:10.1016/j.tics.2006.08.008 PMID:16949331

Decety, J. (2010). The neurodevelopment of empathy in humans. *Developmental Neuroscience*, *32*(4), 257–267. doi:10.1159/000317771 PMID:20805682

Decety, J., & Meyer, M. (2008). From emotion resonance to empathic understanding: A social developmental neuroscience account. *Development and Psychopathology*, *20*(4), 1053–1080. doi:10.1017/S0954579408000503 PMID:18838031

discovery Official Website. (2021). Retrieved from https://www.5discovery.com/en/conflict-management/

Glue Official Website. (2019). Retrieved from https://glue.work/

Hickman, L., & Akdere, M. (2017). Exploring Virtual Reality for Developing Soft-Skills in STEM Education. *2017 7th World Engineering Education Forum (WEEF)*, 461-465. 10.1109/WEEF.2017.8467037

Hussien, A., Tucker, M., Cotgrave, A., Al-Khafajiy, M., & Shamsa, T. B. (2019). Optimizing Project Delivery through Augmented Reality and Agile Methodologies. *2019 12th International Conference on Developments in eSystems Engineering (DeSE)*, 1006-1013, 10.1109/DeSE.2019.00187

McLeod, H. (2018). *Virtual Reality in Leadership Development.* Retrieved from https://www.aacsb.edu/blog/2018/august/virtual-reality-in-leadership-development

Newcastle University Official Website. (2019). *Ground-breaking conflict simulator helps train budding nurses.* Retrieved from https://www.newcastle.edu.au/newsroom/featured/ground-breaking-conflict-simulator-helps-train-budding-nurses

Nvidia HoloDeck Official Website. (2021). Retrieved from https://www.nvidia.com/en-us/design-visualization/technologies/holodeck/

Oculus Business. (2020). *Building empathy to enhance hospitality.* Retrieved from https://business.oculus.com/case-studies/hilton/

Putman, P. G. (2012). *Virtual Simulation in Leadership Development Training: the Impact of Learning Styles and Conflict Management Tactics on Adult Learner Performance.* ETD Archive. Retrieved from https://engagedscholarship.csuohio.edu/etdarchive/249

Singer, T., & Lamm, C. (2009). The social neuroscience of empathy. *Annals of the New York Academy of Sciences, 1156*(1), 81–96. doi:10.1111/j.1749-6632.2009.04418.x PMID:19338504

Spatial Official Website. (2016). Retrieved from https://spatial.io/

KEY TERMS AND DEFINITIONS

Augmented Reality: The AR (augmented reality) technology is relatively recent and constantly evolving. It is the representation of an altered reality in which, to the normal reality perceived by our senses, artificial and virtual information is superimposed, that is a series of information to be superimposed on what the eyes see.

Extended Reality: Extended reality (XR) technology is an umbrella term that includes virtual reality (VR), augmented reality (AR), and mixed reality (MR).

Mixed Reality: Mixed reality is any technology that combines real elements with virtual elements. This classification follows the famous "reality-virtuality continuum" theorized at the same time by Milgram and Kishino, which shows that there is a spectrum of technologies that goes from pure real reality to pure virtual reality.

Virtual Reality: The VR (virtual reality) is a realistic simulation of a reality that does not exist. It comes from the combination of hardware and software devices that "collaborate" to create a virtual space within which the user can move freely. Access to this digital world is made possible by VR viewers and accessories (not just joypads, but also gloves, shoes and more) developed specifically to interact and "live" within virtual reality. In this way a simulated and three-dimensional world is created which in the eyes (but not only) of the users appears to be real.

Chapter 5
Professional Mindset in Troubled Times:
Creative and Collaborative Sharing in the Epistemic Community

Maria Antonietta Impedovo
ADEF, Aix-Marseille University, France

ABSTRACT

Globally, COVID-19 has stressed social and personal tensions in professional life. This chapter focuses on the networked dimension to highlight the workers need for social connections. Some suggestions are proposed to implement an epistemic community to sustain creative and collaborative professional development in disruptive time. Two points are discussed to scaffold epistemic communities in the organisation: 1) the inter-professional dimension to embrace complex topic and 2) the emotional dimension as resources to embrace professional transformation.

INTRODUCTION

The immediate change imposed by the Covid-19 pandemic forced professional and organisation to act and react with urgency. The rapid switch to remotely working at home or the re-organisation of the workplace have required a rethink of the access to the resources, embracing new temporal and spatial configuration to adapt the working condition. At the same time, Covid-19 showed how issues, resources and dynamics are local and global entangled. An intense exchange of information and the growing connections among people, knowledge, and communities could nourish

DOI: 10.4018/978-1-7998-7126-2.ch005

and transform a professional mindset. Two research questions guide the discussion, both from educational literature and from daily ethnographic observation of social networks about working conditions:

1) Which representation of the working condition during Covid-19 emerges in social networks?
2) How does the engagement in online social networks discussions support a creative individual and collective professional mindset, also in times of crisis?

First, a reflection is proposed on the social network's potentiality for professional development. An explorative analysis is done on the sharing in informal social-networks about the working conditions in Covid-19 To answer, at least in part, to the research questions. The paper then proposes a focus on epistemic communities - defined, according to the classic definition of Haas (1992), as a network of knowledge-based experts who together represent the problems they face, identify various solutions, and evaluate the results. In the last sections, two points are discussed to sustain a creative and collaborative epistemic community: a) the inter-professional dimension to embrace complex topic; b) the emotional dimension as resources to embrace professional transformation.

Professional Mindset in Troubled Times

In response to the COVID-19 pandemic started in 2020, many countries have adopted a broad spectrum of containment measures. Corporations and governments alike have strongly encouraged workers to telecommute where possible.

A recent review of the literature points factors that seem to influence our well-being during the quarantine period negatively: 1) the duration of the quarantine itself; 2) the fear to get infected/spread the infection; 3) feelings of frustration and boredom; 4) inadequate supply capacity; and 5) lack of sufficient/salient information (Brooks et al., 2020). To deal with this situation, institutions and universities worldwide have activated online counselling services (Xiao, Zhang, Kong, Li, & Yang, 2020) or shared guidelines to deal with the immediate emergency and promote psychological support.

An informal and easy way to cope in a stressful situation is the narratives, stories and storytelling. Narratives are pervasive in our existence, survived from the past to the internet revolutions. Bruner (1986) considers narrative (based on linguistic connections) as one of the modalities to make sense of the word in dialogue with the paradigmatic manner (based on distinct categories).

The Professional Social Connections: The Role of the Informal Social Networks

The spreading of COVID-19 in the spring of 2020 has brought more professional to isolation and social frustration. As a reaction, online social communities offered an immediate and easy mediated sense of sociality.

There are 7.7 billion people in the world, of which 3.5 billion are online. Social media platforms are used by one in three people globally (Internet World Stats, 2020). Institutions develop the formal social networks with specific learning objectives; informal social networks have been developed with other primary purposes (such as Twitter, Facebook, Instagram and Snapchat) but adapted to training, given the value for social capital. Professionals use widely using informal social networks like Twitter, Facebook, Instagram and Snapchat (Digital Report, 2020), involved in joint discussions worldwide. From a sociological perspective, social media's use responds to social capital acquisition, considered the set of current or future resources (Bourdieu, 1986). Connections have a positive role within communities of individuals. They facilitate coordination and cooperation for mutual benefit, including exchanging resources and building personal and professional relationships, with implications for psychological well-being (Ellison, Steinfield, & Lamp, 2011). The use of these online communities could contribute to workers professional learning (for a review of the literature on this topic, see the work of Bruguera, Guitert, & Romeu, 2019). Networking is a crucial skill in professional careers, supporting the individual's growth and learning. Networks are relationships and spaces where knowledge could show its utmost potential, both for people and institutions (for applying community of practices in international blended learning, see Ginestié & Impedovo, 2020).

Social media's main advantages for lifelong learning and professional development are sharing resources and ideas, facilitating knowledge exchanges and cultivating different forms of solid or weak relationships. The interactions facilitate the coordination and cooperation for mutual benefit and discussion of resources, personal and professional relationships and implications for psychological well-being. For example, social networks like Facebook that were primarily conceived for socialisation have become oriented towards professional use - a phenomenon called professional Facebooking.

Social networking makes possible a networked community of practice (Wenger-Traynor et al., 2015). A network of practices involves individuals connected by social relationships, whether these relationships are strong or weak. The perspective is anchored in the social perspective of "practices" as the activity that connects individuals in their networks. The networked community of practice emphasises the centrality of relationships, personal interactions and connections between people who share information, valuable resources, and individual or collective professional growth. The implications of networkers are visible in using the leading social networks in professional of the so-called Global South (for a study on teacher professional developing, see Impedovo, Khatoon & Kinley, 2019).

Working Conditions During the Covid-19: Contemporary Digital Narratives

Uncertainty and profound shifts in our social and professional routines during a troubled time - as the Covid-19 - stresses the role of coping, resistance, and connection (Outley et al., 2020) strategies in professional networks. Making sense of pandemic situations involves stories and storytelling to share with peer and co-workers. The narrative sharing in informal online social networks about the Covid-19 is full of pain, and in some way, they are also full of derisions. The narrative process could be done through active discussion in groups and visual messages proposed by the meme's contemporary digital culture. In the following, an analysis is performed about sharing the working condition during the Covid-19: the first example is a post in Rabbit groups. Reddit is a social news platform where registered users can submit content in forums. The note in Figure 1 shows the willingness to share the positive benefits to work at home.

Figure 1.

Post in r/jobs (27-01-2021)

The 316 readers reply to this post. The readers agree with the positive modality shifting in the working setting, sharing details about their professional situation and experiences. As one of many, this example shows the necessity to connect with others working, share, and validate, challenge, and extend personal perspectives.

Another example that shows the deep willing to sharing between workers in troubled time is the meme. Memes about the pandemic have been flourishing from the onset of COVID-19 (Blair, 2020; Zhabotynska, 2020). The memes are considered a more straightforward and briefer unit of online content, associated with ironic perspectives, to hit the readers' attention (Shifman, 2014). The meaning is sometimes opaque because it is contextualised to specific communities and events (Literat & van den Berg, 2019). Memes include intertextuality: indeed, they are ready to open generative connections with other contents (Laineste & Voolaid, 2017), framing news events, attitudes and behaviours (Ross & Rivers, 2017). Here we focus on meme related to the work topic to make evidence about representations of professional conditions during the COVID-19. We aim to explore the meme as an artefact used by workers to share, discuss, connect (also only virtually) with peers and co-workers. Some findings are reported from an analysis of 60 memes collected from Reddit's groups in January 2021. The memes are selected from the group "CoronavirusMemes" and ". The meme included in the analysis has a direct reference to the professional life and the COVID-19. In the following Figure 2, some examples of recurrent meme typology inside the group:

In general, the meme shows: the derision of the changing indications to perform; the paradoxical situation of some job positions heavily exposed to the virus; the immobility of the working situation considering the resistance to the home working

Figure 2.

in many contexts; and the control showed to deal with the new conditions of home working.

These memes show how informal sharing of the "quick" narrative becomes a way to deal with the Covid-19 pandemic – as a unique, unusual and, in some way, absurd situation (MacDonald, 2020). The post and the mems collected reduced evidence about the need for workers to connect, share and make sense of their experiences, open to the informal, emotional and narrative side of the social interactions.

SHARING IN CREATIVE AND COLLABORATIVE EPISTEMIC COMMUNITY

The previous section set the place of the narrative and social interactions for professionals in troubled time- as shown by the Covid-19. As a flexible solution to sustain professional development in crises or difficult times, the Epistemic communities could be implemented as a discursive space in the organisation. Online, in face-to-face or blended modality as the best combining of the material and online settings, the epistemic community could engage the workers in original and creative storytelling.

Below, two points to implementing a professional epistemic community are suggested: first, the inter-collaboration to embrace complex issues; second, recognise the emotional dimension as resources to embrace transformation in a troubled time.

The Inter-Collaboration to Embrace Complex Issues

Covid-19 showed no one person with all the knowledge and skills to understand and solve complex issues and phenomena. It implies the necessity to open a bridge of collaboration with different professional. Collaborating, a different profile of workers inside the same organisations may learn to "see" and interpret task affordances and contextual constraints in new ways. This inter-collaboration could open new mutual understanding and reciprocal guided participation (Rogoff, 2003). The inter-professional collaboration could facilitate the workers to assume different identity positions, helping to reframe and enrich perspectives in a mutual commitment.

The conversation becomes the privileged place to carry out an explicit negotiation and redefinition of shared values, learning how to become a recognised member of one other community (for example, the community of the accountants and the administrative staff's community in the same organisation). Through the others "ways of saying", each member interprets and play with the roles and move in the different skills spheres. The sharing can be superficial and oriented to exchanging information and resources or more engaged in reflexive practices, social presence and critical engagement to generate new knowledge linked with the organisation s' aims.

The community's participatory nature could facilitate forms of so-called relational agency (Edwards, 2017), which involves using knowledge to take engaged action with others. This space of experience can be also a space of tensions and frustrating between them. Indeed, joining in an inter-professional discussion means engaging in an articulated conversation where there is an explicit professional commitment to seek mutual understanding but not necessarily agreement. The challenge is to address the "inner logic" of the interlocutor's speech to make evident the invisible fundamental way to see the others, as suggested by Sfard, 2019. Indeed, all participants are invited to recognise the problem's interests as necessary; even they may disagree on how to address it. For example, an organisation could provide a physical or online discursive space in which one member of the community can (or not) freely decide to join into the discussion, with the explicit recognition of the possible divergent opinions.

The Emotional Dimension as Resources to Embrace Professional Transformation

The COVID-19 opened a negative loop of emotions in the light of personal loss and economic and social struggles due to sanitaria restrictions. In these troubled times, the invitation is to embrace the critical situation: not to deny it but to give open access to the emotional dimension. The workers' invitation is to stay with the trouble by 'learning to be genuinely present (Haraway, 2016, p. 1). One strategy could be to awaken access to the personal (emotional, social, experiential) workers' past as professional resources. It can be a unique background made up of emotional, social, professional experiences (not in line with the main path), developed in voluntary associations, informal setting, family, etc. Indeed, not yet, the workers' personal experiences are evaluated as a precious ally in professional developing and quality practices. In this way, the professionals not cut off their unique specificities but valorise the idiosyncratic emotions and experiences as a valuable resource: the working set became an emotional resonance setting addressed to the collective healing. Esteban-Guitart (2016) develops the funds of identity concepts. It originally mobilises the identities of the minority in educational settings for social justice purposes. Two dimensions are related: the dark funds of identity (Charteris, Thomas & Masters, 2018), defined as the problematic experiences that individuals bring with them to make sense of theoretical concepts; and the second is existential funds of identity (Poole, 2020), defined as the positive and negative experiences that individuals develop and appropriate to express themselves and to grow as human beings. The dark funds of identity and the existential funds of identity could be a prosperous professional resource if recognised, supporting the professional development's emotional side and supporting an authentic growing inside the workplace.

The arts of noticing (Tsing, 2015, pp. 17–25) suggest a fruitful creative and innovative reflection on the experiences. In a difficult or troubled time, the reflexive process on the personal and collective professional development could be done in a "diffractive" prospective - like the lights' effect on a diamond. This "diffraction perspective" refers to a patter where the meanings are shaped in a lived way, shifting the gaze from individuals to human and to the more-than-human entanglements. (Haraway, 2016). This concept invites to reframe the reflective methods from a static representation of reality to an interconnected relationship with social, economic, environmental issues. According to Hill (2017), reflective methods are often related to a static representation of reality, which is assumed to be pre-existing and stable. Instead, the diffractive reflections "illuminate the fluid and ever-evolving process of world-making in which phenomena are constituted through their material entanglements" (p.3).

Adopting this perspective means to legitimate the workers' expression of emotions - instead of hiding them. The emotional and personal experiences become a helpful resource for collectives learning and development. So, in a troubled time, an organisation's leadership management is invited to be aware of the professional's emotions and experiences.

CONCLUSION

This paper aimed to engage the organisation in implementing and developing epistemic communities as a physical, online, or blended space to engage in multimodal narrative storytelling in a troubled time.

Enterprises could take advantage of existing social networks -or develop internal solutions - to sustain a sense of social connections in a troubling time, reducing the sense of isolation due to the home working. The networking modality, the opportunities to exchange with different professionals about complex issues and the ability to express and valorise personal and emotional resources become a unique opportunity for professional development. Also, digital analysis of these spaces could give the organisation some insights to support individual or collective quality professional development. So, the organisation could improve the workers' well-being to deal with the critical and challenging time increasing collaboration and creativity. Indeed, a multimodal interaction is facilitated by innovative technology (like the use of avatars in 3D Virtual Worlds), making it accessible for the workers to experiments with tools and solutions.

A networked, inter-professional, and inclusive community can engage professionals to find and discuss solutions, generating new knowledge and solutions to the advantage of personal and collective professional development and well-being.

In summary, advice for practitioners in their organisations are:

- COVID-19 have stressed some social and personal tensions in professional life globally;
- The networked dimension enables and maintains social connections;
- The implementation of an epistemic community could help to sustain professional development in troubled time;
- The inter-professional dimension support to deal with complex issues;
- The emotional dimension, if recognised and valorised, could become a resource to embrace professional transformation.
- Emergent technology could support a multi-modality of social and communicative interaction (virtual, augmented, synchronic, etc.).

In conclusion, the epistemic communities' implementation in an organisation means to support the single and the collectivity in a deep sense of belonging to the organisation.

REFERENCES

Blair, O. (2021, January 27). 125 coronavirus memes to get you through self-isolation and social distancing. *Elle*. https://bit.ly/3gkvblr

Bourdieu, P. (1986). Habitus, code et codification. *Actes de la Recherche en Sciences Sociales*, *64*(1), 40–44. doi:10.3406/arss.1986.2335

Brooks, S. K., Webster, R. K., Smith, L. E., Woodland, L., Wessely, S., Greenberg, N., & Rubin, G. J. (2020). The psychological impact of quarantine and how to reduce it: A rapid review of the evidence. *Lancet*, *395*(10227), 912–920. Advance online publication. doi:10.1016/S0140-6736(20)30460-8 PMID:32112714

Bruguera, C., Guitert, M., & Romeu, T. (2019). Social media and professional development: A systematic review. *Research in Learning Technology*, *27*(0). Advance online publication. doi:10.25304/rlt.v27.2286

Bruner, J. (1986). *Actual minds, possible worlds*. Harvard University Press. doi:10.4159/9780674029019

Charteris, J., Thomas, E., & Masters, Y. (2018). Funds of identity in education: Acknowledging the life experiences of first-year tertiary students. *Teacher Educator*, *53*(1), 6–20. doi:10.1080/08878730.2017.1367057

Edwards, A. (Ed.). (2017). *Working relationally in and across practices: A cultural-historical approach to collaboration*. Cambridge University Press. doi:10.1017/9781316275184

Ellison, N. B., Steinfield, C., & Lampe, C. (2011). Connection strategies: Social capital implications of Facebook-enabled communication practices. *New Media & Society*, *13*(6), 873–892. doi:10.1177/1461444810385389

Ginestie, J., Impedovo, M., (2020). International Teachers Professional Developing: Blended Learning between Europe and Asia. *Voprosy obrazovaniya, 2,* 114–127.

Haas, P. (1992). Introduction: epistemic communities and international policy coordination. *International Organization. Cambridge Journals, 46*(1), 1–35. https://www.jstor.org/stable/2706951

Haraway, D. (2016). Staying with the Trouble: Making Kin in the Chthulucene. Duke University Press. doi:10.1215/9780822373780

Hill, C. M. (2017). More-than-reflective practice: Becoming a diffractive practitioner. *Teacher Learning and Professional Development*, *2*(1), 1–17.

Impedovo, M. A., & Khatoon, S. Kinley, (2019). Global South teacher educators in the digital landscape: implications for professional learning. *Research on Education and Media, 11*(2). doi:10.2478/rem-2019-0018

Internet World Stats. (2020). https://www.internetworldstats.com/stats.htm

Laineste, L., & Voolaid, P. (2017). Laughing across borders: Intertextuality of internet memes. *The European Journal of Humour Research*, *4*(4), 26–49. doi:10.7592/EJHR2016.4.4.laineste

Literat, I., & van den Berg, S. (2019). Buy memes low, sell memes high: Vernacular criticism and collective negotiations of value on Reddit's Meme Economy. *Information Communication and Society*, *22*(2), 232–249. doi:10.1080/1369118X.2017.1366540

MacDonald, S. (2020). What do you (really) meme? Pandemic memes as social, political repositories. *Leisure Sciences*. Advance online publication. doi:10.1080/01490400.2020.1773995

Outley, C., Bowen, S., & Pinckney, H. (2020). Laughing while black: Resistance, coping and the use of humour as a pandemic pastime among blacks. *Leisure Sciences*. Advance online publication. doi:10.1080/01490400.2020.1774449

Poole, A. (2020). Re-theorising the funds of identity concept from the perspective of subjectivity. *Culture and Psychology*, *26*(3), 401–416. doi:10.1177/1354067X19839070

ReportD. (2020). https://wearesocial.com/digital-2020

Rogoff, B. (2003). *The cultural nature of human development*. Oxford university press.

Ross, A. S., & Rivers, D. J. (2017). Digital cultures of political participation: Internet memes and the discursive delegitimisation of the 2016 US Presidential candidates. *Discourse. Context & Media*, *16*, 1–11. doi:10.1016/j.dcm.2017.01.001

Sfard, A. (2019). Learning, discursive fault lines, and dialogic engagement. In N. Mercer, R. Wegerif, & L. Major (Eds.), *The Routledge International Handbook of Research on Dialogic Education*. doi:10.4324/9780429441677-9

Shifman, L. (2014). *Memes in digital culture*. The MIT Press.

Tsing, A. L. (2015). *The mushroom at the end of the world: On the possibility of life in capitalist ruins*. Princeton University Press.

Wenger-Traynor, E., Fenton-O'Creevy, H. S., Kubiak, C., & Wenger-Traynor, B. (2015). *Learning in Landscapes of Practice: Boundaries, Identity and Knowledgeability*. Practice-Based Learning.

Xiao, H., Zhang, Y., Kong, D., Li, S., & Yang, N. (2020). The Effects of Social Support on Sleep Quality of Medical Staff Treating Patients with Coronavirus Disease 2019 (COVID-19) in January and February 2020 in China. *Medical Science Monitor*, *26*. Advance online publication. doi:10.12659/MSM.923549 PMID:32132521

Zhabotynska, S. (2020). The narrative multimedia concept: An algorithm for the analysis (internet-memes about COVID-19). *Cognition, Communication, Discourse*, *20*, 92–117.

Chapter 6
Learning From the Past:
Intelligence in Decision-Making Processes During Prisoners of War and Hostage Crises

Immacolata Carcarino
Alma Mater Studiorum, University of Bologna, Italy

ABSTRACT

Learning from past experiences should be considered vital in every decision-making process. It becomes even more important when lessons could save lives. Past experiences could indeed support the development of survival courses and reintegration processes for military and civilian personnel deployed in hostile environments. Moreover, the integration of lessons from the past and basic techniques could be useful for anyone working, or simply traveling, in areas where the risk of isolation and/or kidnapping is considered very high. During an isolation event, every aspect should be considered from support to families during captivity to the reintegration into everyday life. Governments, NGOs, and international organizations should look at the past and learn how to establish an effective system to repatriate and reintegrate their representatives.

INTRODUCTION

The numbers of prisoners in modern conflicts are impressive. During WWI between 7.000.000 and 8.500.00 persons from all nations were captured: over 420.000 by the French, almost 525.000 by the British and something in that range by the Italians, 1.500.000 by the Russians, 1.700.000 by the Germans, and 2.000.000 by the Austrians.

DOI: 10.4018/978-1-7998-7126-2.ch006

The estimates for WWII are inflated due to the collapse of the German state and the surrender of Japan. Entire armies of the defeated powers became prisoners during the last six months of 1945, raising the highest estimates for that period to 35.000.000 (Davis, 1977).

The reaction of the public opinion to the news describing the kidnapping of military and civilian personnel in operations can affect the political decision-making process. Those men and those women are sons, daughters, fathers, mothers, brothers, sisters, husbands, and wives, before having a role in operations. Watching them on the news or in captors' frightening videos can touch the soul of any human being that can easily imagine his/her own son or brother in the same situation, causing a strong empathy for the victims and their families. The sustainment of operations, to domestic and foreign policies can be undermined by the weight of those, but it is not the only risk that a nation has to manage. Indeed, prisoners of war (POWs) can be exploited by the enemies to collect precious operational information. Those procedures are forbidden by the Laws of Armed Conflicts, but history has taught us that few nations really cope with those regulations. Since ancient times, POWs have been a product of wars and important sources for the enemies to learn more about their adversaries. Focusing the attention on contemporary history, with modern conflicts increasing the mobility of troops, asymmetric threats superseding traditional warfare and non-state actors turning into ruthless captors and, the number of POWs, and victims of captivity, surged enormously.

At their return, prisoners can be welcomed as heroes or as criminals, depending on the circumstances, as experienced for example during the Vietnam war. They can fear the opinion of their countries and even experience post-traumatic stress disorder (PTSD). They are fragile, but they can provide an amount of vital information that should be collected promptly with effective techniques from a specialized team. For all the above-mentioned reasons, POWs enter a Reintegration phase, starting once they reach friendly facilities. From that moment on, an articulated process, divided in distinctive phases, will guide them on their way back to normal. It may take just hours, but it may also last for months, sometimes years, depending on POW's reaction upon return and his/her physical and psychological conditions.

In this chapter, POWs intelligence will be divided in two distinctive phases: intelligence in captivity and intelligence after captivity. The first phase includes all actions carried out to acquire information from enemies, but also the actions taken by POW's country to prevent social media and journalists from spreading sensitive details about the operation, the hostage and his/her family. The second step consists of all the intelligence activities conducted after the POW is back under friendly control (either released or successfully escaped), entering in a Reintegration process. All the valuable information collected in this phase could potentially enter the intelligence cycle to be then processed and disseminated through the communities of interest.

Nothing can be left to chance when dealing with POWs, from both sides of the barricades, but we should never forget that before being "information bearers", they are human beings, deserving respect and healthy living conditions.

During the recent lockdowns, we have all been prisoners of war, a different war. Our captor is invisible, but it seized our lives with evil cruelty. Each one of use experienced isolation from family and friends, many lost beloved ones without having the opportunity to tell them goodbye, but we all are striving to survive. As we wait to be released and come back to life, writing of POWs in this time of crisis and suffering may arouse strong feelings and their stories can be now seen with different eyes.

POWs sacrifice their lives, with death or perpetual physical and psychological sufferings. They could be considered heroes or traitors, friends, or foes, means or ends, but they are undoubtedly victims that are to be protected and treated humanely. Therefore, this work looks at POWs as victims and resources, scrutinizing past events and providing recommendations for the future in an attempt to raise awareness on a topic repeatedly faded into oblivion.

Background

For Carl von Clausewitz, war is "a paradoxical trinity–composed of primordial violence, hatred, and enmity, which are to be regarded as a blind natural force; of the play of chance and probability within which the creative spirit is free to roam; and of its element of subordination, as an instrument of policy which makes it subject to reason alone."

The battlefield may become a peaceful heaven compared to the hell POWs experience once captured, even though international agreements forbid reprisals against them. The regulations concerning POWs were first detailed in The Hague Conventions and, later in Geneva Conventions, that became a symbol of the fight against the atrocities WWII POWs went through. It was impossible to forget what happened during that conflict and the aim of those Conventions was to avoid other barbarities in the future. Those wounds were still bleeding when the third Geneva Convention was signed in 1949. When they finally healed, the memories of all the pain disappeared. History repeated itself in Korea and Vietnam. The lesson was not learned.

Rules concerning POWs treatment are based on ancient military codes of conduct, dating back to chivalry times, and aim to secure a humane and respectful treatment to all POWs in all circumstances, with the only distinction based on their ranks, if military personnel. Since the nineteenth century, nations started to agree on the importance of regulating the treatment if POWs during hostilities. Prior to the nineteenth century, prisoners of war had no protections under the law. Captors punished, enslaved, or even murdered their prisoners. Several events during the last half of the nineteenth century were instrumental in the creation of international humanitarian law concerning prisoners of war. Korea and Vietnam wars had indeed a significant impact on international law due to the evil treatments experienced by American and Communist prisoners of war. After 1975, when Vietnam War was finally over, the Geneva Conventions were revised again with Additional Protocols I and II in 1977. The Protocols addressed issues concerning the differences and similarities between international and non-international armed conflicts.

The nature of conflicts is not static, it evolves with time. Thus, International humanitarian law must adapt to changing operational environments to provide deployed personnel with the adequate protection. In modern unpredictable asymmetric warfare, it may be challenging to accomplish this task, but intelligence support can make it possible.

History demonstrated that POWs are victims, but they also have important and various roles in many crucial domains for both the detaining power and their country of origin. As discussed in previous chapters, they are sources for intelligence, during and after captivity. Furthermore, captives can act as ignition for political debates, influencing public opinion and decision-makers. They may also affect the economy and the culture of their captors. The impact on these last two spheres is less relevant, but it demonstrates that POWs can be considered valuable resources in various fields.

INTELLIGENCE AND DECISION-MAKING PROCESSES DURING PRISONERS OF WAR AND HOSTAGE CRISIS

The Role of Intelligence

Intelligence was not recently invented. As long as states have existed, governments have made all possible efforts to collect information about potential, internal or external, adversaries. Based on their knowledge and available techniques, secret services or individual agents have gathered information to support policy-makers in foreign and domestic affairs or to military forces during operations.

Nowadays, intelligence analysts have a plethora of technologies to make reports more reliable and useful, especially with the increasing globalization of international relations. As a consequence, in the last century all countries experienced the rise of well-funded and well-equipped intelligence agencies capable of coordinating the various sources and processing an ever increasing amount of raw data. Among all the different types of intelligence, there is still one that favors the most ancient sources of information: human beings.

Human intelligence (HUMINT) is an essential part of any efficient security or intelligence system. Despite all the valuable information technologies can provide, human beings remain valuable sources. Considered risky and difficult to handle, HUMINT requires skillful personnel and dedicated training.

During conflicts, interrogations of POWs can be defined also as HUMINT activities (Tobia 2015, p. 71), as enemies are using human sources to obtain information.

In war, intelligence on enemy forces is critical for commanders' decision-making process. In some conflicts, when enemy air superiority hinders the use of air reconnaissance assets or, in case of lack of sources for signal intelligence, POWs may be the only specialized intelligence sources available. The process of exploiting POWs is known as interrogation. The main obstacle an interrogator can meet is the communication with the prisoners. Language barriers, for example, could affect the conduct of the interrogation and making harder, if not impossible, to collect information. In addition, a complex process for the organization and the planning of the interrogations is required. Interrogation is important, but not an easy task for the captor.

Prisoners of War Intelligence is defined as "information acquired during the skillful interrogation of prisoners or the monitoring of their conversations with fellow prisoners or stool pigeons that can be of critical importance" (West, 2015, p. 281).

In POWs intelligence, the same hindrances usually experienced in HUMINT can be encountered. They are related to the willingness or reluctance of individuals to betray their families, countries or other allegiances that may prove difficult to control and liable to work to their own agenda, to fabricate information, or to become double agents.

POWs Intelligence

POWs Intelligence can be categorized in two phases: Intelligence in captivity and Intelligence after captivity. The distinction between the two phases is not limited to the different circumstances that captives can experience. The desired results and the methods used to gather information can differentiate the two stages of the Intelligence process concerning POWs.

Undoubtedly, POWs represent precious sources of information. Indeed, they could provide details about military maneuvers, weapons, force composition, but also important information about sociopolitical situation in their countries that could be exploited by hostile forces for their purposes. This phase can be defined as Intelligence in captivity, during which captors take advantage of their position to obtain information from POWs. Meanwhile, the country of origin of the prisoners must apply any possible procedures to protect POWs and their families, especially in recent conflicts, due to the over exposition to social media. It is to be remembered that the survival of POWs should always be considered within the realm of national interest and security. More important is the fact that the prisoners are still citizens of their country, and as they are presumably coming back, their well-being and morale must be vital. That a prisoner of war camp is a safe place to relax and to be away from the dangerous battlefield is a myth. The majority of those who are fortunate enough to be alive at the conclusion of the war will have external and internal scars for all the rest of their lives.

The aim of the first phase is to become cognizant about the adverse party: any method could be used to achieve that goal. Meanwhile, friendly intelligence services will protect the POWs from any possible threat, intended or not, for instance exposition to social media, information leak about their next of kin, sharing of sensitive details related to rescue operations. Historical analysis reveals that, in general, interrogation conducted closer to the time and place of capture are more effective. In those critical moments, it is possible to collect information of current intelligence value. On the other hand, interrogations at later stages have historical and strategic values. Intelligence in captivity results in being highly time-sensitive. Historical information is related to events that occurred more than twenty-four hours earlier than the time for the interrogation. Current information is indeed related to events occurred within the last twenty-four hours and could be on strength, locations, and present missions of enemy units. Specific questions and tailored report forms can be used during interrogation.

How the captors collect information is not a mystery. There are many techniques, most of which quite questionable, to convince prisoners to disclose classified operational information. For Machiavelli, "the end justifies the means", but is the end more valuable than human life? And how effective are the means?

For the UN Convention against Torture and Other Cruel, Inhuman or Degrading Treatment or Punishment, torture "means any act by which severe pain or suffering, whether physical or mental, is intentionally inflicted on a person for such purposes as obtaining from him or a third person information or a confession, punishing him for an act he or a third person has committed or is suspected of having committed, or intimidating or coercing him or a third person, or for any reason based on discrimination of any kind, when such pain or suffering is inflicted by or at the instigation of or with the consent or acquiescence of a public official or other person acting in an official capacity." Notwithstanding the unanimous condemnation, torture was, and still is, one of the main information-gathering tools worldwide. The effectiveness of inhumane techniques to extort confessions of any kind had always been questioned. Napoleon Bonaparte wrote in his "On the subject of torture, in a letter to Louis Alexandre Berthier", dated 11 November 1798 "the barbarous custom of having men beaten who are suspected of having important secrets to reveal must be abolished. It has always been recognized that this way of interrogating men by putting them to torture produces nothing worthwhile." It is legitimate to doubt the reliability of information acquired under torture. To stop the inhumane treatment, a prisoner could say anything, true or false. Neuroscientific researches demonstrated the effects of typical stressors experienced in captivity, such as waterboarding, hunger or sleep deprivation.

The second phase starts once the prisoner is handed over to his country or rescued. During this seemingly endless period, the former prisoner is assisted, medically and psychologically, to be reintegrated into society and in his/her job. All the findings will be analyzed and will support the intelligence cycle of the specific theatre in which he/she was hostage.

In addition to their tactical and intelligence value, prisoners are useful as subjects, objects and instruments of propaganda. Charges of mistreatment are probably the most effective theme of propaganda concerning prisoners of war. The second world war had its Bataan Death March, and countless war crimes charges registered at Nuremberg and Tokyo trials. War crimes allegations also dominated the interminable peace negotiations of the Korean and Vietnamese wars.

Historically, being prisoner of war has meant that service members could experience brutality, coercion, loneliness, and many other forms of deprivation. These experiences can accentuate human dependence on captors, achieving an utmost degree of exploitation. One of the major worries plaguing military personnel, should they become prisoner of war, is that of the relationship between military responsibility and personal survival. Survival in prisoner of war camps may involve instinctual rather than rational behavior. The greatest service military personnel can render as prisoners is to remain true to themselves and to serve with silence and honour in the military way. Captivity could be indeed seen as an extension of the battlefield where the prisoners carry on the struggle with the only weapons remaining in their possession: faith, brainpower, and courage. POWs have the obligation to continue to help their countries and their countries expect their soldiers to give their lives, no matters where this call comes to them, may it be on the battlefield or in a POW compound in some strange land.

The immediate and lifelong effects of these exploitations are documented and cannot be underestimated. Military personnel captured and detained as POWs present significantly high rates of emotional and physical trauma, posttraumatic stress disorder (PTSD) and other mental health conditions. During World War II roughly half of the service members captured in Germany and Japan developed PTSD which persisted throughout their lifetime and had extremely high mortality rates and cognitive difficulties. Moreover, between 88 and 96% of Korean War POWs experienced a mental health condition related to their captivity. Some of these problems seem to be related to severe malnutrition and mistreatment often experienced by POWs and enlisted among the worst war crimes; those who lost 35% or more of their body weight during captivity have had the greatest degree of verbal and visual learning and memory deficits. The causes of these aftermaths were determined by variety of complex and chronic stressors to which POWs were exposed. Some of the typical physical stressors were exposure to temperature extremes (hot and cold), indigenous infections, environmental toxins, denial of medical care, dehydration, and nutritional deprivation. While isolation (both solitary confinement and controlled communications), uncertainty about captivity duration, unpredictability of captor behavior, lengthy interrogations, and the ever-present specter of death generated hopelessness, helplessness, fear, anxiety, guilt, and shame. In addition to physical and psychological stressors, torture was frequently used by the enemy. Torture is an experience that cuts across both the physical and psychological dimensions, involving both passive (e.g., forced standing, forced kneeling, and sensory deprivation) and active (e.g., beatings, suffocations, mock executions, and various methods of joint dislocation) techniques (Kennedy, Moore, 2010).

Trained to Survive

To contrast enemies' techniques and procedures, an adequate training is required. In the US, the earliest survival schools focused on the use of life rafts, and provided the admonition, if captured, to disclose only the "Big Four" (name, rank, service number, and date of birth). Following WWII, when the Air Force was created in 1947, basic survival schools were set up in Nome, Alaska; Thule, Greenland; and Goose Bay, Labrador. Since the primary Air Force mission at that time was defending Alaska and preventing attacks over the North Pole, these schools were subsequently created to prepare service members for cold weather environments and taught such skills as building makeshift airstrips for rescue.

It was the Korean conflict, however, that dramatically changed the focus of the survival schools. Although the Korean War has been referred to as the "forgotten war" (fought between WWII and the Vietnam War), this description marginalizes the physical and psychological injuries suffered by many of these POWs. Following the Korean War, 21 service members agreed to stay in Korea, having signed false confessions. Many interrogation experts and consultants believe that these confessions were the result of physical and psychological torture. Following these events, former POWs and senior military leaders began to take a long and serious look at how to better prepare service men and women in survival training. Survival, evasion, resistance, and escape (SERE) training schools in their current form were developed by a working group established by President Eisenhower and implemented by the US Air Force in 1961.

The innovation was the inauguration of military "torture schools" in which roleplaying soldiers prepared for "the abrupt transition from American life to Commie prisoner life." (Carruthers, 2009). "We must do everything we can psychologically to prepare our soldiers for the torments of these heathen Red foes," insisted General Matthew B. Ridgway, who had commanded the UN forces in Korea. At Stead Air Force Base in Lemmons Valley, Nevada, airmen encountered a regime in which "hunger, pain and fatigue" were part of the training. So too was being "forced to strip, to hear lies and insults about their personality, race, religion, national origin or physical characteristics." (Ibidem)

When Newsweek broke the story about Americans confining one another in 120-degree temperatures in sweatboxes, instigating mock "death marches," and using live electric shocks to their peers, many commentators were alarmed. The Washington Post and the Saturday Review protested, referring to those schools as "school for sadists," noting: "Brutality is like a bullet; you don't shoot a man to prepare him for war. And when you degrade a man and humiliate and damage him the end result is measured in the damage it does to him, not in the supposed 'training' he receives". After all these negative comments from the general public, the Pentagon changed the program. The Air Force announced that the course was only on a volunteer basis and after three months it was announced that the trainee would have merely watched instructors using Communist interrogation techniques on one another, following a suggestion made by Senator Lyndon B. Johnson.

After the Vietnam POWs returned in 1972, several of them aided their SERE schools by teaching students about their experiences with torture, lengthy interrogations, threats of execution, disease, physical injuries, communications with fellow POWs and, most important, the means to keep hope alive.

Back to Life

Prisoners of war can come back to life, or in Dante's word, "they could get out and see the stars". Whether they are rescued, released, or successfully escaped from captivity, POWs can go back to their countries. The recovery or escape of the prisoner is not the final act. There is still a long way to his/her return to a normal life. The truth is that not all POWs come back home. Some may die, others may decide to stay. As strange as it may sound, it did happen in the past. In past conflicts, some of them were also "reeducated" and eventually "converted" to their captors' beliefs.

All persons who have been captured, and successfully recovered, will surely require medical and/or psychological care, but can also hold valuable information of immediate tactical Intelligence that may lead to the recovery of other isolated personnel or to the successful engagement of the enemy. They may also be able to share lessons learned regarding avoidance of capture, coping strategies and survival. Learning from the past has never been as beneficial as in this case. All the events of captivity experienced in history helped improving military techniques, tactics, and procedures in operations, thanks to all the people that directly experienced them.

"There is a lot of pain in the world that people have, but physical pain is not the worst. The worst pain is the pain of loss, when you have nothing, no family, no home, no ties, you have nothing—that is the worst pain in the world". These words were spoken by an American captive returned from Vietnam in 1973. At that time, public opinion learned for the first time about the treatment of POWs and how they were assisted to their return to life after captivity. From that moment on, US Department of Defense has implemented a system to guide former POWs from captivity to ordinary life.

The emotional trauma of captivity is likely to leave a residue of psychic scar tissue that never altogether heals (Gober, 2005). Prisoners of war are at high risk for the development of psychological symptoms, including shame, depression, PTSD, and other anxiety disorders. Although this topic may seem remote to civilians, the possibility of being taken hostage occurs in many terrorist scenarios and is a real possibility for those in the Foreign Service and others working abroad.

Personnel considered at high risk for capture may, and should, receive specialized training, known as survival, evasion, resistance, and escape training. In this extremely intense course, service members and civilians are "captured" and subjected to mock interrogation, sensory deprivation, and other humiliations. In part, this represents primary prevention against the development of psychological symptoms in the event of capture. In the wake of the capture of non-trained personnel in the Iraq War, there has been discussion of expanding the training to more service members. Moreover, some international and nongovernmental organizations have also developed training for their personnel. It is important to underline that the reactions of military and civilian personnel are different. These differences are based on different personality characteristics, different motivations, and different expectations. As a result, the subsequent handling after recovery requires techniques tailored to the specific category in order to be successful.

Understanding the phases of captivity is helpful in planning the response. Immediately after capture, there may be a real possibility of being killed by the captors. There may be many relocations, long painful marches through freezing cold or deadly hot weather, and passage through the hands of many captors. In World War II, the Korean War, and the Vietnam War, this initial turbulent phase was followed by a long period of relative sensory deprivation, with years spent in enemy POW camps. Malnutrition and chronic medical diseases can be exacerbating factors. Many of these potential problems associated with being a POW can be alleviated through careful reintegration into society. In the reunion phase, emphasis is given to the idea of "decompression." This allows the returning POW or hostage to receive medical and psychological care before being subjected to the media lights or the meetings with family members, friends, and local politicians. The released hostage may present signs of disorientation, hypersensitivity, disorganized thought, aversion to physical contact, to name a few. Prisoners are persons who experienced deprivations, alterations of circadian rhythms, sustained and persistent stress, denial of personal hygiene. They are fragile and must be properly treated, protecting them from the external world and media and providing a calm and quite environment.

This phase of a recovery operation is called Reintegration. It is a national responsibility, and it is designed to debrief the recovered isolated personnel, provide medical and mental health treatments, and assist the recovered personnel and their next of kin in resuming their professional and private lives. The aims of this process are, but not limited to: return the recovered person, physically and emotionally healthy to duty; gather evidences to prosecute captors; allow recovered personnel to understand the isolating event fully and make them gain a sense of mission completion, helping their units by providing precious information about the enemy.

Any single detail of their stories may be relevant, so the Reintegration teams have to take record of everything. Even apparently useless information, for example voices and languages heard in the place where they were held could provide proofs of living of other hostages still in captivity. The types of questions asked during debriefings include, for example: how were they isolated; what are enemy tactics and procedures; who is with them; identities and behaviors of their enemies that lead to the isolation event; and if captured, the identity and behaviors of their captors.

A prisoner needs medical stabilization, but also to regain the ability to predict and control. It is beneficial for them to tell their story repeatedly in a healthy manner, having their reactions and emotions normalized. After the initial phases of mission completion and time-sensitive tactical debriefing, medical treatments are needed. These stages of Reintegration can be lengthy and are dependent on the reaction of the POW. The role of families here are vital. Next of kin can be help the prisoner, but they need to be supported by specialists to guide them through this path.

In 2014, after five years in Taliban captivity, US Sgt Bowe Bergdahl went through Reintegration after long captivity. During a news conference, his father compared his son's experience to a deep-sea diver returning to the surface as it needs to be done slowly. "If he comes up too fast," he said, "it could kill him." Reintegration takes time and both POWs and families are to be assisted through the whole process. A POW not reintegrated is a person left behind in captivity.

POWs in Political Decision-Making Processes

We all take decisions, every day and in every single moment of our lives. We decide what to eat, what to wear or where to go on holidays. In our very personal decision-making processes, we identify the problem, consider, and evaluate alternatives, choose the solution that best fulfils our personal requirements, and allocate the resources we need to implement, under the influence of our values and knowledge. External factors can intervene, and external voices (family, friends, social media) can lead or divert us along our way to the final decision. What happens when a government or an organization must take a decision? Notwithstanding the different relevance of the decisions, the steps to follow are almost identical, with more variables and more influencing factors. Leaders often face inconceivable dilemmas alone, in a group or in a coalition, during their decision journey. The subject of their scrutiny may be at a national or international level and their choices may affect the lives of millions of people. In addition, environmental factors, i.e. time and information constraints, stress, risk or ambiguity, and cognitive biases may affect the decision.

During WWII, and even more during the Cold War, politicians and military leaders relied on intelligence as a vital tool for their decision-making processes, using propaganda and covert operations to achieve their goals. Intelligence in war can be problematic due to the difficulties in understanding, analysing and predicting the enemies. In recent campaigns in Iraq and Afghanistan intelligence was not successful in supporting decision-makers. US forces had little information about tribal dynamics and the role of neighbouring countries. In Rupert Smith's words, "in order to understand operations amongst the people, and to capture their will, we must first understand 'the people'. The people are an entity but not a monolithic block. They form entities based on family, tribe, nation, ethnicity, religion, ideology, state, profession, skill, trade and interests of many different kinds." (Smith, 2012) Do the prisoners of war have a role in these processes? In recent and past conflicts, events involving POWs catalysed the attention of politicians and general public.

The *Pueblo* crisis in 1968 is a great example of decision-making, even though this event did not occur in a "war", so prisoners would be better defined as hostages in this case. The whole decision-making process was indeed carried out to avoid another conflict in an already unstable area.

The *Pueblo* was a U.S. Navy vessel on an intelligence mission off the coast of North Korea to monitor its radio and electronic transmissions. On 23 January 1968, North Korean forces attached the ship, killing one crew member, wounding four and capturing the rest of the crew, 82 persons. After 335 days in captivity, US hostages were released.

President Lyndon B. Johnson opted for negotiations and diplomacy, excluding any military options, whether a rescue mission or an attack. In both cases, there would probably have been more victims than recovered personnel. Moreover, the domestic protests against the war in Vietnam were massive and US leadership was not in the position to start another conflict: another war was clearly not an option. Many individuals in US were more favourable to a military action against North Korea, including three future Presidents, Nixon, Ford, and Reagan.

Using analogical reasoning, decision-makers recalled all past events that could have some similarities with the crisis they were handling, even though they often referred to that crisis as unprecedented and peculiar. It was indeed the first time in recent history that a US vessel surrendered in peacetime. Browsing through the pages of US history, the first documented surrender of a US ship happened in 1807, when the USS Chesapeake gave up to the British without a fight. The last seized vessel was the USS Manning Light in 1863, captured by Confederated Forces. From 1968, the year of the Pueblo Crisis, one had to go back 161 years for a direct peacetime precedent, and at least 105 years for a similar incident in wartime.

Looking for analogies in hostage crisis was also hard because the few precedents were mostly many decades before. The Barbary Pirates affair was, for example, the first example of hostage crisis for the US. In 1803, 307 Americans on board the frigate Philadelphia were captured in Tripoli harbour and released after two years of captivity. A century later, in 1904, US had to deal with another hostage crisis, the Raisuli affair in Morocco. President Theodore Roosevelt obtained the release of the hostages through a mixture of diplomacy and military threats. Some members of the Congress looked at those historical events, but President Johnson was guided by other lessons like the 1962 Cuban missile crisis, widely regarded as an outstanding success and a model for sound decision-making and the 1964 Tonkin Gulf incident, in which a rapid response to flawed intelligence about a spy ship had helped lead the US into extensive military involvement in Vietnam. This last event was crucial for the decision-making process in Pueblo crisis. President Johnson was perfectly aware of what happened the last time he responded rapidly with incomplete intelligence information. That time he authorized a military response, despite the conflict and contradictory information received before the final authorization.

In Defense Secretary Robert McNamara's word, the first analogy is explained "I think we need a Cuban missile crisis approach to this, and goddamn it, we ought to get locked in a room and you ought to keep us there, insist we stay there, until we come up with answers to three questions: what was the Korean objective, why did they do it; secondly, what are they going to do now - blackmail us, let it go; and thirdly, what should we do now?"

USS *Pueblo* crew was exploited for propaganda purposes by North Koreans. The hostages reacted to captivity and secretly dismantle their captors 'propaganda. In all photos and video, they showed a middle finger, explaining that it was a good luck sign. It became a part of their anti-propaganda campaign until North Koreans learned what that finger really meant, when Time magazine published a photo of the men, writing in the caption that "three of the crewmen have managed to use the medium for a message, furtively getting off the US hand signal of obscene derisiveness and contempt." The reaction of the captors was merciless, but luckily the crew was released soon.

North Koreans also used documents stolen during the seizure of the vessel to show to the international community the espionage activities conducted by the US, in this document: "*Les Actes D'Agression Declares de L'Imperialisme U.S. Contre Le Peuple Coreen: Les matériaux concernant les actes d'agression du Pueblo, vaisseau-espion armé de l'armée d'agression de l'imperialisme américain ayant fait profondément intrusion dans les eaux territoriales de la République Populaire Démocratique de Corée*".

USS *Pueblo* crisis became itself an analogy when Cambodia captured the USS *Mayaguez* and held the crew hostage. President Ford launched a recovery mission that resulted in a failure, causing the death of 41 Marines.

While the Marines were disembarking from the CH-53 helicopters, the hostages had been already unexpectedly released. 41 lives were lost in vain, but the reason of the release is still unknown.

Pueblo and *Mayaguez* incidents were used as analogies to find a solution to the hostage crisis in Iran in 1979. The US embassy in Tehran became the target of the wave of criticism against US during the revolution by Islamic fundamentalists. On November 4, 1979, Iranian students seized the embassy and captured 63 American diplomats. The Iranians held the American diplomats hostage for 444 days, during which President Carter's conduct of foreign policy was undermined and the decision-making process appeared weak and vacillating.

This event is also a good example of groupthink syndrome, characterized by the "narrowing of vision" (Mintz, DeRouen 2010), leading to decisions taken without confrontations, being based on overconfidence and similar views. Groupthink can be avoided if leaders let other voices speak during the decision process. It is exactly what happened during the Cuban Missile Crisis, when President Kennedy told his executive committee to seek for alternatives and to think "out of the box". A devil's advocate, in that specific case Secretary of Defense Robert Mc Namara played that role, can also help finding unbiased options. During the Iran crisis, the only devil's advocate was excluded from the decision-making process and the results were not successful.

Notwithstanding the accuses of the Iranians, US had limited intelligence capabilities at that time in Iran, due to a substantially cut back of the CIA's HUMINT capabilities that Carter championed to counter the negative public perception of the agency. In a 1979 confidential memo, CIA official Paul Henze shared his thoughts in the crisis and confirmed the irony of that moment: "It is supremely ironic that we should stand accused of so much espionage out of our Embassy in Tehran when we have done so little. What modest efforts have been in recent months will no doubt be shattered by what has just happened. But we cannot simply close up and forget Iran. We have had so little luck with the intelligence problem here that a special task-force effort to gather intelligence on Iran through all possible channels outside of Iran and through friendly governments inside of Iran needs to be made".

Secretary of State Cyrus Vance and National Security Advisor Zbigniew Brzezinski had divergent opinions on many foreign policy issues, including the handling of the Iran crisis. Vance had a more moderate approach than Brzezinski, preferring negotiations to aggressive stances. As a result, Vance was kept out of the decision-making process in a classic groupthink style, when the secret rescue mission in Iran "Operation Eagle Claw", promoted by Brzezinski, was authorized, and planned. Vance was against that decision, that was taken in his absence. He resigned in protest in April 1980. His position proved correct, as the mission ended up as failure that cost President Carter the 1980 reelection. There were many lessons to learn from that debacle, one of which contributed to developments in areas such as military preparedness. In the final report of the Special Operations Review Group, Admiral Holloway confirmed that "[the operation] was risky and we knew it, but it had a good chance of success and America had the courage to try." The planning and training phase was long, lasting 6 months and included several exercises for the crews, due to the complex environment and a plethora of uncertainties the planners had to deal with, including the lack of reliable intelligence on the location of the hostages and the turbulent political circumstances. On the evening of April 24, 1980, eight RH-53 helicopters took off from the aircraft carrier Nimitz to the refueling site Desert one. Due to bad weather conditions and helicopters' failures, only five helicopters were made it to the refueling site. The minimum number of assets to continue the mission was six, so the mission was aborted by the President, but unlucky events occurred while repositioning, causing the death of eight crew members and the loss of some recovery vehicles.

Operation "Eagle Claw" left many lessons for future operations. The Special Operations Review Group gave some recommendations, such as the implementation of intelligence for hostage release operations, the importance of joint training and evaluation, and to improve American Special Forces capabilities, in general. The hostages were released the following year as a result of diplomatic negotiations.

Another example of groupthink dates back to the Vietnam War and is related to another failed rescue mission. On 21 November 1970, an American task force of 56 men was involved in one of the most audacious operations in Vietnam. Based on some information suggesting that 70 American POWs were held in Son Tay camp, the recovery vehicles took off from Thailand in a night operation under radio silence. The operation was rehearsed 170 times and planned with extensive intelligence and logistical support, but nothing went as planned. As soon as the rescue team landed in the camp, it was evident that it was the wrong location.

The only available information about North Korea at that time were photographic reconnaissance, and a specialized unit, the 1127th Field Activities Group, based in

Fort Belvoir, Virginia, tasked with locating active POW camps. On 9 May 1970, the analysts identified possible camp activities at Son Tay. They discovered POW uniforms arranged in the courtyard shaping the letters "SAR" and rocks forming the letter "K", which were both codes used for search and rescue. Another pile of rocks conveyed a message in Morse Code, informing that there were at least six men in the camp that were going to die if not rescued on time. The US started planning the operation, even though intelligence confirming the presence of POWs in Son Tay was scarce. "Vietnam was a war of politics in which political expediency took priority over military necessity" (Amidon, 2005) and the Son Tay Raid was planned and executed when the United States was negotiating a conclusion to the war. The withdrawal of ground forces had already been completed in the south, and one of the few remaining tasks was the return of American POWs. In this milieu of this critical balance, the Son Tay raid represented a pivotal point in Vietnam War history. The operation, as it was planned, should have proved the military superiority of American Forces to North Vietnam in an effort of convincing them to accept negotiations. Due to the high strategic impact of the operation, both President Nixon and his National Security Advisor Henry Kissinger were briefed on the raid. Kissinger was doubtful about the risk of having more POWs in case of failure, but military planners predicted a 95 to 97 per cent assurance of success. During the briefings, military authorities never revealed that the presence of POWs in Son Tay had never been confirmed. President Nixon was enthusiastic and "lapped it up like an eightyear-old at his first cowboy movie." When one of the military officers mentioned the possibility to abort the mission in case of enemies in the area, Nixon protested "Damn…let's not let that happen. I want this thing to go." (Ibidem) The authorization for the mission was given without any hesitation, but the result was not as expected. The first rescue team landed in the empty Son Tay prison, while the second team mistakenly landed 400 meters away and had to fight a bunch of enemies, before being able to join the rest of the team for the withdrawal.

The groupthink syndrome in Son Tay can be identified in both White House and Pentagon decisions. Notwithstanding the lack of reliable and clear intelligence, the Pentagon recommended the mission "go" to President Nixon. Moreover, there were few successful rescue missions in US military history to use for analogies. Typical symptoms of groupthink can be found in Son Tay decision-making process: illusion of invulnerability, belief in the inherent morality of the group, collective rationalization, out-group stereotypes, self-censorship, illusion of unanimity, direct pressures on dissenters, and self-appointed mind guards. All people involved in the planning and decision-making of the raid in Son Tay were firmly convinced of the success of the operation, despite the lack of solid intelligence and hiding all possible sign of failure. After all the consequences of previous decisions concerning Vietnam,

President Nixon's staff shielded him from the aftermaths of a failure at Son Tay, acting like perfect mind guards, even concealing the existence of possible risks.

The Son Tay raid was not a debacle. It was a well-executed mission, that showed the ability of US military in Special Operations. Even though no prisoners were rescued during the raid, it had some positive effects. As a consequence of the operation, North Vietnam gathered all POWs in fewer locations to prevent other rescue attempts. American POWs had more opportunities to communicate. The raid also positively affected their morale, demonstrating that their nation had not forgotten them. The most successful result of the Son Tay raid was restoring POWs' hopes.

Prisoners of War or War of Prisoners?

During the WWII, Winston Churchill had to "host" some "unexpected guests" (Jackson, 2010). Following the Geneva etiquette, the guests were treated with respect and occupied a special place at the British intelligence table. The main concern of the British PM as a host was to avoid reprisals against British "guests" in German POW camps. Churchill and his military advisors knew exactly that any action against German POWs could have been reproduced on their military personnel, as already experienced in October 1942. Following the raid in Dieppe in October 1942, the Germans announced that they would have chained the hands of all British soldiers, because this was the treatment German POWs had suffered from their enemies. The reprisal and counter-reprisal went on, with accusations and threats, until the Swiss Government intervened as a neutral mediator to stop the pantomime.

Both parties had to deal with turncoats and deserters. The first category included all POWs that found a "vocation", true or coercive, and started fighting with their former enemies. Russians POWs in German hands discovered themselves particularly devoted to their opponents' cause and joined them in combat. Their choice was driven by the harsh treatments Russians POWs were suffering, and by knowing that Stalin considered prisoners as missed in action. Despised by their captors and repudiated by their country of origin, their only option was to go back to the battlefield, fighting under another flag. On the contrary, deserters were more difficult to exploit. Not always reliable as stool pigeons or as sources of information, they were considered more keen than useful. Though strongly willing to help the enemy, they randomly gathered information without understanding the context. Along with deserters, another category of POWs difficult to handle was "the diseased". Ill captives needed special care and were unproductive. Despite the costs for the detaining power, POWs with disease could also represent a health risk for the local population. Some of the "unexpected guests" shipped to Britain, mainly Italians captured in Africa, were already sick at their arrival. Italian POWs were affected by malaria, typhus, and other minor pathologies, that forced British authorities to find urgent solutions to avoid the spread of potentially lethal diseases among the civilians. The remedial action was to concentrate sick prisoners in designated areas, following rigorous procedures to mitigate the risks for British citizens. Never was a guest as unwanted as the Italians were in Britain.

Welcomed or not, POWs can repay their captors' hospitality with their precious information. Should a list of favourite guests be produced, U-boat crews occupy the highest position. British government knew they were valuable sources and was lucky enough to have them as the first prisoners to arrive in the country. German captives revealed tactics, procedures and routes that were vital for future successes of the Allies. The main contribution was for the disclosure of the use of "Enigma". The cipher machine, employed by Germans to encrypt messages, was composed by sets of "wheels" that could be used in billions of combinations, and was on board U-boats carried them on board. After many attempts to find an intact machine and useful instructions, the British finally succeeded with "Operation Primrose" on 9 May 1941, when British vessels captured a U-boat, taking all documents and an Enigma before the crew could destroy it. The classified documentation, codes, instructions, and machine were sent to Bletchley Park, the location where British Intelligence officers and mathematicians, led by Alan Turing were studying the Enigma codes. The final pieces could finally complete the jigsaw and decipher German communications. For some historians, that British advantage shortened the war by two years. If it really made the war end earlier or not, the capture of German sailors and the subsequent revelation of Enigma's secrets provided the Allies with crucial information that paved their way to the victory.

During the Cold War, in the United States, Chinese and North Koreans POWs were exploited to obtain a different victory: they should have been an example to follow for their communist comrades back home. Over the course of the early cold war, the Truman and Eisenhower administrations in turn developed policies to promote defection, and champion POWs' right to refuse repatriation. Government officials strongly emphasized that thousands of Chinese and North Korean freedom lovers had been saved from totalitarian slavery. The main objective of those policies was indeed to encourage communist troops in any future war to surrender at the earliest opportunity, bearing in mind the infinite opportunities of a new life in another country. The fear of massive desertions would have made the Kremlin avoid future wars. "From now on, the Red Armies will be less dependable as tools of aggression. We have increased the prospect of peace and added to the security of our nation," (Carruthers, 2009) were the words of Secretary of State John Foster Dulles in September 1953. Prisoners of war became the "weapons" in the war of prisoners. The Americans had found an option to win the Cold War without the use of military force, but they also had to win American public opinion's consensus. Surveys conducted in 1952 demonstrated that although a growing number of respondents favored refusal to return reluctant communist POWs, a majority would have supported the policy only if it had not jeopardized the return of American POWs (Ibidem). News of brainwashing and atrocities perpetrated against American captives in Korea generated waves of criticism and fear. As effective as it could have been, the desertion policy was not fully endorsed by public opinion. Americans wanted their POWs back.

Nowadays, in an overconnected world, POWs can be victims and resources as never before. The power of new technologies and internet can make captives more vulnerable, accompanying them in a new dimension of public curiosity.

The Article 13 of the 1949 Third Geneva Convention relative to the treatment of POWs (GCIII) prescribes that "prisoners of war must at all times be protected, particularly against acts of violence or intimidation and against insults and public curiosity." The recent International Committee of the Red Cross's updated Commentary on GCIII (2020) present a new insight into the protection of POWs. Images and videos can be as humiliating and harmful as in the past. The physical exposure has now become virtual, with POWs and hostages marching in cyber parades on social media and online contents, available 24/7 to a broad and uncontrolled audience. Their faces, names and personal lives are presented indiscriminately, jeopardizing the safety of their families, and of the prisoners themselves, once released. Duty to report notwithstanding, it is vital to control and securely store all data and contents which depict POWs. Should the pictures be broadcasted, methods such as blurring, pixelating or obscuring faces and name tags must be used in order to protect prisoners' identities. Internet and TV shows could become new prisons for captives. They are victims of the morbid curiosity of improvised keyboard warriors, ready to fire piercing comments from their living room command posts.

This new dimension of captivity can also provide new opportunities to exploit POWs. Captives have always been precious resources in wartime. For example, the advantages expected from the capture of military personnel can be economic. Capturing enemies prevents them from inflicting damages to the captor and reduces the total fund of manpower and skill available to adversaries. The economic advantages are indeed marginal and could include the acquisition of weapons and equipment, use of personal property the captives may have with them, plus a share of the relief supplies sent to the prisoners from abroad and diverted by commerce or theft into local economies. The International Red Cross sent great numbers of parcels to American and British POWs during both world wars that enabled Germans to evade some of the maintenance costs. Indeed, the main economic contribution of POWs derives from their labour during captivity.

Along with the economic benefits, POWs contributed to the enrichment of captors' culture and arts in past conflicts. POWs can be interned, tortured, and humiliated, but they will always remain human beings with their own stories, memories, feelings, and passions. Wherever they go, they bring their hearts with them. Diaries, newspapers, letters, and drawings provide valuable historical information on ordinary life in a camp. The human side of a prisoner can be seen in their works. Captives produced objects using what they could find in the camp. With creativity, they could turn a toothbrush handle into a chess piece or metal tins into a model steam engine. There were also satiric drawings and funny cartoons. They could find strength in creating comic strips and make fun of their fellow prisoners or of their captors. If they could still draw and create something, their soul was still alive, and the body simply had to follow. Those rare occasions represent a moment of connection between enemies and prisoners, during which distances and hostilities are neutralized by the power of art and imagination.

POWs could also be used to divert the attention of public opinion from other events. A famous example is the capture of the American POW Jessica Lynch, a former US Army soldier. She was part of a maintenance unit that was victim of an ambush in Iraq in March 2003. After nine days in an Iraqi hospital, she was rescued and returned home safely. Her capture was extensively reported and discussed in all American news: her blonde hair and blue eyes were everywhere, representing the typical young girl next-door sent to the front line to fight for her country, risking her life. The rescue mission was entirely filmed, providing a Hollywood flavour to the military operation. Some newspapers presented Jessica Lynch as a brave warrior who fought with all her strengths against the enemies. The former soldier has always told the truth about the ambush, taking the distance from those headlines, calling herself a survivor rather than a heroine. "I'm an American soldier, too" were her words to the rescue team, invigorating the patriotic spirit in a controversial war. The news of the rescue appeared few days after President Bush announced the start of military operations in Iraq. All media were interested in the young American soldier's story, a mixture of fable and military heroism, creating an American icon catching the attention in those crucial days.

That event also demonstrated that anyone could become a POW and should be prepared for it, no exclusions applied, paving the way to new policies concerning survival training.

CONCLUSION

In the words of Winston Churchill, a POW is described as "a man who tries to kill you and fails, and then asks you not to kill him." In those words the essence of POWs' dichotomy is revealed: executioner and victim. This chapter has followed a different approach, leaving the executioner in the battlefield, and highlighting the human side.

Prisoners of war are protected by international humanitarian law, a set of rules that should prevent detaining powers from torturing and exploiting them. Dating back to chivalry times, international humanitarian law had been updated throughout the centuries. Nevertheless, history has taught us that law can be violated. From brainstorming to death marches, modern conflicts are a perfect representation of how law can be ignored. Whether it was for propaganda or strategic objectives, POWs were victims of the most barbarous crimes a man can endure. Surviving for them is even worse than dying. Many POWs will live with nightmares as lifelong friends.

History has also taught us that all detaining powers take advantage of their role and exploit prisoners to achieve their objectives. POWs can be used for propaganda or work for their captors or become stool pigeons. They can also become "students" in reeducation programs, too often more focused on brainwashing than teaching. In either case, the results were not brilliant: coercion is not a guarantee of success. Prisoners usually try to find the best option to avoid sufferings and pain. They can indeed prefer confession to resistance, indoctrination to captivity. Therefore, reliability can be undermined. Confessions can become a collection of lies and indoctrination an impromptu divine call that will fade away as soon as they come back home.

The most successful operations that could really achieve some military advantages required long and complex intelligence activities, i.e. Operation Primrose in 1941. In POWs intelligence, the risk of failure is extremely high, especially when both captors and captives are not trained. If an interrogator ignores the right questions to be asked or how to collect them, the results are doubtful. At the same time, if a prisoner is unable to resist stressors, he/she may compromise the political and military objectives of his/her country.

The United States learned the lesson after the war in Korea and set up a system to prepare military personnel in case of capture in future conflicts. They also developed a code of conduct to guide them. Those Korean lessons were tested in Vietnam where American POWs could react and survive the long captivity. Education and training are always important but become vital when someone is deployed in hostile environments.

This chapter analyses POWs, but the importance of education and training should be addressed to all personnel deployed in areas with a high risk of isolation and capture, i.e. people working for NGOs, international organizations or civilian companies. Military lessons could support the implementation of civilian and governmental pre-deployment and awareness courses. In other words, a "toolbox" with essential survival notions could be supplied to anyone working abroad. Furthermore, the preparation phase becomes even more crucial when POWs or hostages are released. Despite the different protocols applied to military and civilian captives, reintegration should be implemented in either case. For example, Italy has recently repatriated some hostages and the effects on public opinion and on returnees themselves were shattering. The morbid curiosity of journalists, the improvised strategic messages, and the subsequent sterile debates are the fatal errors to avoid when dealing with hostages. In future operational environments, civilians will be even more exposed to the risk of isolation and kidnapping. Therefore, a remedial action should be taken before it is too late.

Blinders and bias have never been good mentors. If military and civilian entities continue to proceed on parallel directions, they will never meet and will never share precious lessons that could potentially save human lives. POWs can teach us many lessons, but only if we are willing to learn them.

Lessons, but also memories of astonishing humanity. Victims become valuable resources, demonstrating that flowers can bloom even in the driest soil. Prisoners of war left tangible signs of their presence in many countries, sharing their knowledge and skills with others. In rare case, they arrived as enemies and left, or even remained, as friends.

Churchill was partially right. It is true that POWs tried to kill, but they also saved lives sharing their experiences and lessons learned.

REFERENCES

Amidon, M. (2005). *Groupthink, politics, and the decision to attempt the Son Tay rescue*. Air War Coll Maxwell AFB AL.

Carruthers, S. L. (2009). *Cold war captives: Imprisonment, escape, and brainwashing*. University of California Press. doi:10.1525/9780520944794

Davis, G. H. (1977). Prisoners of war in twentieth-century war economies. *Journal of Contemporary History, 12*(4), 623–634. doi:10.1177/002200947701200401

Gober, S. J. (2005). *The reintegration of military families following long term separation*. Arizona State Univ Tempe. doi:10.1037/e452872006-001

Howard, M., & Paret, P. (1984). *Carl von Clausewitz On war*. Princeton University Press.

Jackson, S. (2010). *Churchill's unexpected guests: Prisoners of war in britain in world war II*. The History Press.

Kennedy, C., & Moore, J. (2010). *Military neuropsychology*. Springer Publishing Company.

Mintz, A., & DeRouen, K. Jr. (2010). *Understanding foreign policy decision making*. Cambridge University Press. doi:10.1017/CBO9780511757761

O'Mara, S. (2018). The captive brain: Torture and the neuroscience of humane interrogation. *QJM: Monthly Journal of the Association of Physicians*, *111*(2), 73–78. doi:10.1093/qjmed/hcx252

Smith, R. (2012). The utility of force: The art of war in the modern world. Penguin UK.

Tobia, S. (2015). Close encounters: The role of culture in interrogation and questioning in the second world war. *Rúbrica Contemporánea*, *4*(8), 69. doi:10.5565/rev/rubrica.94

West, N. (2014). *Historical dictionary of british intelligence*. Scarecrow Press.

Chapter 7
Symbioses as an Alternative to Master/Slave for Artificial Intelligence

John Thomas Riley
The Big Moon Dig, USA

ABSTRACT

Human society is facing enormous problems this century as result of our climate crisis. These problems include sea level rise and the loss of farming capability. Society will need all the new tools it can develop to address these problems. Artificial intelligence with deep learning is one of these powerful tools, and it is new. Exactly how it will be used has not been determined. The current approach to the human/AI interface is referred to as master/slave. The human simply tells the AI what to do. This arrangement has many problems, and replacing it has been suggested. One possible new arrangement is a human/AI symbiosis. This would require a long-term relationship between a specific human and a specific AI. A novel, Born to Storms, exploring this arrangement is discussed at length.

INTRODUCTION

Artificial Intelligence [AI] is a new tool that society needs in order to address the great problems of the 21st century. Most human and AI interactions are now based on the concept that the human will give orders and the AI will follow them, an arrangement that has long been known as "Master/Slave." Recently, this term has been considered historically insensitive and alternative terms have been proposed.

DOI: 10.4018/978-1-7998-7126-2.ch007

An alternative to the Master/Slave concept is needed beyond just a change in name. This chapter envisions the development of a symbiotic relationship between humans and their machines as a strategic new direction for the human/machine relationship, formally a Symbiosis.

Our society now faces a number of grave crises. One of these is Our Climate Crisis, which will define the 21st century. Society will need all the tools it can obtain to address the resulting problems. AIs and Deep Learning are two of the most attractive new tools that are now available for this effort but will require both adaptation and application.

An example of the symbiotic arrangement from a novel for young adults is discussed below. The work, *Born to Storms,* is a novel about a young human/AI symbiont working toward a career addressing the myriad problems of Our Climate Crisis.

As a starting point for this approach, an understanding of what artificial intelligence is, and the recent breakthrough that is propelling it forward, Deep Learning, are needed.

DEFINITION OF ARTIFICIAL INTELLIGENCE

An Artificial Intelligence [AI] is a machine that can perform some useful function normally expected of humans and not machines. These functions include facial recognition and discovering trends in very large data sets. People now commonly interact with an AI that makes suggestions for their purchases or that finds spam in their emails. Sometimes an AI is the control system for a robot, but it does not have to be. More commonly, AIs are interacted with through a simple voice or a text box.

Most AI interact in some way with human beings. The AI may be limited to hearing a voice, or seeing a picture, and may have some limited physical capability, such as the money slot of an automatic teller. Other AIs interact less directly with people by inputting a large data set developed from human activity and returning the tools needed for humans to understand analysis such as charts and graphs.

A.I. is a new tool that will be needed to address the great problems of the 21st century, especially Our Climate Crisis. People and machines can generate a great volume of data, but that data requires sophisticated analysis to support the many difficult actions that humans must take to address real problems and opportunities.

A recent advance in technology, Deep Learning, may make this tool available in a timely manner.

Figure 1. Diagram of Deep Learning Matrix

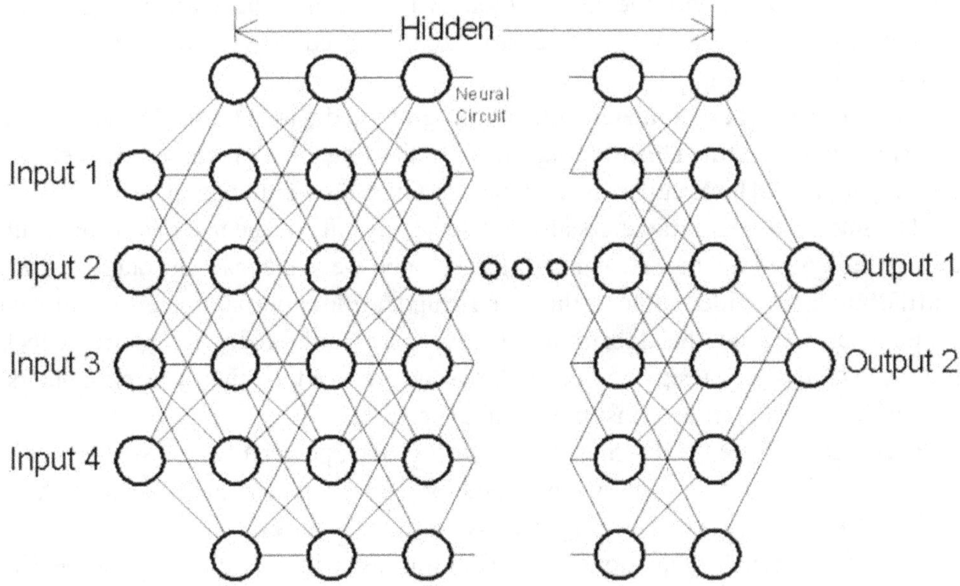

DEFINITION OF DEEP LEARNING

Deep Learning is a field of computer science that develops the algorithms that allow AI to produce a useful result from a data set and to learn from the data (Russel, 2015; Kelleher, 2017; Burkov, 2019). It is usually applied to large data sets to produce such results as facial recognition, sales assistance, and business trends.

Deep Learning is one of a number of mathematical procedures that might be applied to generate this learning product. For Deep Learning, the computer memory is organized as a matrix of cells that functions as a biological nerve cell, the neuron. To formally qualify as Deep Learning, there must be at least three internal or hidden columns in addition to the input and output cells.

In some applications, there can be thousands of rows and thousands of columns and require substantial time and expense for training. Considerable effort is now being made to reduce these costs and speed up the training process.

Figure 1 shows a matrix of self-contained neuron-like circuits. Each row receives information from many of the cells in the row before it and provides information to the cells in the row after it. The first row provides input to the system and the last row provides output. Each cell has a number of internal parameters, hyperparameters, that support the calculation of its output given its inputs. The recalculation of these hyperparameters constitutes learning for the system.

Note that the neural circuits connect strongly with the column before and the column after but not with the cells in their own columns. The exact configuration of these many connections can be customized to speed up specific tasks such as face recognition.

The internal hyperparameters must be programmed with a target data set for the algorithm to be useful. This training involves dividing an appreciable data set into two major parts. The first part is then hand evaluated by a human expert.

The internal hyperparameters are first set at any value. The test data is then run though the system and the result noted. The error between the machine output values and the human-provided values is then used to update the hyperparameters. With each training data element, the algorithm must sweep forward through the matrix, check the result, and then sweep back through the matrix to adjust the hyperparameters. This training can be time-consuming and expensive.

Fortunately, since the internal cells are in a matrix, powerful and fast algorithms for matrix arithmetic, originally developed for games, are available. Without these tools, the learning process would probably be too slow to be of value.

Often the internal connections and the training algorithm are chosen for a specific task, like facial recognition. Having multiple Deep Learning chip sets each dedicated to specific tasks, rather than one generalized set, may be needed to demonstrate capabilities that are more useful for a complex AI.

If Deep Learning is such a breakthrough, why has society not already widely embraced it?

ROLLERCOASTER RIDE

AIs have gone through an extended rollercoaster ride in the popular imagination, with many ups and downs (Kelleher, 2017). Each time there is an advancement [like winning at checkers, then chess, and then Go] the advancement is overhyped in the popular press. This is then followed by a long period of no reported progress. This ride has failed to build realistic expectations for AIs with the general public.

Sometimes a technical advancement has direct practical application, like voice recognition or sales. In these situations, there is little need for complex technical explanations for the public if the result is easy to use and of help to them. People easily accept help in finding a product they want and often do not even recognize the salesperson as an AI.

Other times an AI project is technically sweet; then technical people will pour time and resources into the project. That is what they do. Only a few of these projects will lead directly to public breakthroughs, but many more will advance the technology and become buried in later breakthroughs. AI has enjoyed this type of behind-the-scenes support for decades and is now seen as the most likely field that a wannabe technical wizard should choose for a career.

A new upturn in this rollercoaster ride is expected soon as custom hardware chip sets for Deep Learning become widely available and algorithms to drive them are developed. AI are then expected to become fast, low power, and everywhere.

But having a new and powerful tool is not enough; society has to apply that tool to the great problems it faces. What then is the current relationship between society and AIs?

DEFINITION OF MASTER/SLAVE

Clearly, most people expect an AI to take orders from a human operator and not show any signs of rebellion. A formal definition of its process [Called Master/Slave (Technical)] is:

- One unit is primary – This master unit is the center hub of a network of machines. This node usually is human programmable and subject to shut down.
- Other units are secondary – These slave units are outliers in the network and accept instruction and data from the central hub.

The term "Master/Slave" was first used for timekeeping equipment designed specifically for safe operation of the railroads and appears in related literature by 1904. By the time early computer equipment was developed in the 1960s and 1970s, the term was well established. It was specifically used when one central computer was outfitted with numerous data input terminals (Kidder, 1981). The term fell somewhat out of use by the public with the widespread introduction of individual personal computers. Later, it was strongly reestablished for the equipment that hosts the Internet.

The term Master/Slave (Sexual) is also used in a sexual context. This is **not** the sense that this term is used in this chapter.

Aside from that same-name problem, why is the term Master/Slave so controversial?

CONTROVERSY OVER TERM

The term "Master/Slave" has been labeled racist and offensive by some movements trying to achieve social equality (CNN, 2003; Seattle PI, 2004; Wired, 2020). Social pressure has been applied to technology companies to change their terminology.

Within a few years of these proposed changes, the "Black Lives Matter" movement eclipsed the earlier efforts.

Proposed alternatives include *replica, standby, secondary*, or *follower* instead of *slave* in code, in conversations, and in product descriptions. That said, such a longstanding technical term is difficult to change, especially as the offense was not intentional and the alternative terms may sound awkward to established users. This change in terminology has not caught on.

What is needed is an alternative to the process of Master/Slave itself and not just a change of name. This calls for a change not just of labels but also in foundational concepts.

PROBLEMS WITH MASTER/SLAVE

The Master/Slave process has built-in problems. There is no reliable way to ensure that who is the master and who is the slave cannot flip. This is a strong societal concern and has been depicted in several modern movies.

The expectation of the general population is that the capabilities of the AI will continue to improve through technological innovation while humans can only improve through much slower training over generations. One side outpacing the other could easily flip a Master/Slave relationship, much to the original master's disadvantage.

Is the history of human slavery a driver of this problem?

DIFFICULT HISTORY OF SLAVERY

It is quite difficult to enslave a population and still get productive work out of its members.

If enslaved in their own land, the enslaved people tend to run away easily and often. Despite the claims that Christopher Columbus made in letters to Queen Isabella, Native Americans made poor slaves for this very reason. They may be at work today, but they can be many miles away by tomorrow morning. They could then reliably live off their own land and stay out of sight.

Slavery in many forms has been a fixture of human society to provide controlled manual labor through most of history. Its use required that the master class be eternally vigilant against slave rebellions. Two famous examples from the history of major failures of slavery are the Roman Spartacus Rebellion and the Antebellum South.

The Roman Empire initiated nearly continuous wars. Its economy was dependent on the booty brought back and the slaves captured from the losing side. One rebellion was led by the gladiator Spartacus. An army of escaped slaves ravaged the large farms they once worked on for two full years, marching up and down the boot and winning several major battles. They were finally put down when enough Roman legions were sent to seriously outnumber them and the pirates that were to take the ex-slaves to Sicily were bought off.

There were a number of serious slave rebellions in the Antebellum South. Slavery there was a particularly severe form, chattel; people were property. This harsh form was seldom used throughout history. The rebellion incidents generated so much fear in the white population that the entire society was warped by the effort to prevent rebellion. It is hard to argue that your society is tranquil and Christian if a lower class in held in its place with whipping and your society includes posses of heavily armed slave catchers riding through the night. The largest of the rebellions occurred near the end of the period when John Brown led a raid on the Federal arsenal at Harper's Ferry. His plan was to steal the military firearms manufactured there and move through the South arming any escaped slaves that would rally to his banner. An alarm of fear ran throughout the South until John Brown was caught and hung.

The American Civil War was set up by the introduction of two technologies, the cotton gin and a gun with interchangeable parts; both were invented by the same man, Eli Whitney. The last forms of slavery were finally displaced by the introduction of new technologies, but pockets drag on to this day in hidden pockets. A civil war might have been avoided if the new technology of steam had been introduced more quickly. Such an introduction would have generated major social disruptions no matter when it came.

Slavery in the United States finally became obsolete with the introduction of powered machinery, particularly in agriculture. This process started with steam but became universal with the introduction of diesel power. The Romans had industrial-level waterpower and steam toys, but never built a steam power plant. Their progress to a modern power economy was blocked by their dependence on slavery.

There is no reason to argue that AI slavery will not end any better than human slavery. There is good reason to worry that the effort to keep AIs slaves could warp human society into a bad configuration and result in major restrictions to the development and application of AIs.

Given this history, how than has the relationship between society and AIs been portrayed in popular entertainment?

AI SLAVE REBELLIONS IN THE CINEMA

The fear of an AI slave rebellion is clear enough in the minds of the public to support several major series of movies and television shows. Example of such franchises include *Terminator*, *Battlestar Galactica*, etc. Typically, robots, with AI minds, rebel and wreak havoc in human society before being improbably put down by humans.

Clearly, AI rebellion is a deep-seated fear in the human mind. Considerable effort is therefore justified in addressing this problem, even if the specific problem itself, a major AI slave rebellion like in *Terminator*, is unlikely in the extreme.

Are there alternatives available to Master/Slave? Can these alternatives be presented in popular media that society's young people will connect with? Some new concept is required for the core value of such a project. One good possibility is symbiosis.

DEFINITION OF SYMBIOSIS

A Symbiosis is an established relationship between two organisms. In weak symbiosis, each member benefits from the association but can survive without the other. In strong symbiosis, neither member can even reproduce without the other.

Well-known examples of symbiosis include the lichens that are made up of a fungus and a plant. The fungus can recover minerals from weathering rock and the plant can harness sunlight for energy.

Less well known are the symbioses between virtually every plant crop humans depend upon for food [wheat, barley, etc.] and the microorganisms in rich soils. The root systems of food plants cannot draw up enough minerals without the help of an entire ecosystem of fungus, worms, and bacteria in the soil. It is this ecosystem that makes the difference between pale sand and deep, rich soil.

Some relationships are one sided; one member is a parasite. The relationship helps one member but hurts the other. These are common [mosquitoes, tapeworm]. This arrangement needs to be avoided.

Human beings are already dependent on many natural symbioses. Can humans develop a symbiosis involving humans and AIs as an alternative to Master/Slave?

Can we even build a true human/machine symbiosis? Would such a symbiosis be of value to human society?

To answer these questions, we must first consider some realistic possibilities for AI capacities in the near future. Whether AIs will be able to think is a good place to start.

AI CONSCIOUSNESS

By definition, Cognitive Science is the precise and testable theories of the human mind (Russel, 2014). This is the science that looks into human consciousness. As of this writing, technological development is a very long way from building an AI with testable human consciousness, and the effort to build one is not presently a major research effort.

Consequently, full, human-style consciousness in AIs is neither necessary nor likely to happen in the near future. This human trait is simply not needed in AI to function in society nor for the proposed human/AI symbiosis.

AIs are extremely unlikely to develop human consciousness, but they may develop an identifiable AI consciousness. In the same sense that whales have whale consciousness, AIs could develop AI consciousness that is distinctly **not** human consciousness yet still is of great value to human society.

In fact, the point of the symbiosis is to support each participant in doing the things that each does best while covering the areas where the partner is weak. Their capabilities need to complement each other, not match exactly.

If a precisely reproduced human mind was thought to be needed, what is the value of such a mind, and why has this capability been the goal of AI developers for so long?

VALUE OF A HUMANIOD MIND

It could be argued that an AI with a near-humanoid mind would be able to fit more easily into human society. For example, it could provide oral answers to oral questions that people could easily understand. Many of the qualities of the human mind, like the will to dominate others, are not, however, useful in AIs and would have to be eliminated or suppressed.

Commercially, the most cost-effective use of a human-like AI mind would be to fool people, like customers, so that real people would not have to be hired. This is not necessarily a good result for society as a whole, even if it is of commercial value.

One alternative is that the AI mind be a complement to the human mind without being an exact copy. It must provide useful answers and actions and not waste people's time with irrelevancies. It must be an asset to human society but present no actual danger.

The concept of a human/AI symbiosis addresses this problem when the human member of the pair leads in interactions with other people. Guidance in such interactions could also make these an important training experience for the AI.

If mimicking the humanoid mind is a valued goal, why not a humanoid body too?

VALUE OF A HUMANIOD BODY

A human body would let the robot use tools, machines, and facilities designed specifically for humans. Although this presentation is very common in popular movies, it has not yet been achieved as an integral unit in reality.

Such a humanoid robot would also have a very large carbon footprint. It would require a very high level of manufacture and materials shipped from all over the world.

Although the size of the electronics, and therefore its carbon footprint, has dropped by several orders of magnitude, known as Moore's Law, the physical parts of a robot, the bones and muscles, do not fit this law. The best improvement possible in the foreseeable future is only a factor of two or three.

It is therefore unlikely that humanoid robots will be available to help young people address the problems of the 21st century. Therefore, humanoid robots are unlikely to be early candidates for a human/AI symbiosis.

If an exact copy of a mind or body is not needed, how close should these copies come to resembling their human counterparts?

THE UNCANNY VALLEY

A robot or an AI avatar may either look like a machine or look like a believable human. They may not look like a poorly formed human; such presentations are seen by many people as a sick person and are subsequently rejected. This rule is called the Uncanny Valley and applies to both full-fledged robots and to human like images on the screen.

The humanlike quality is grafted against acceptance by customers, the line first forms a peak for obvious robots; dips forming a valley; and then rises again for very human like presentations. This dip was quite unexpected, hence the name.

The AI member of the symbiont must either appear as a very good representation of a human or appear as some other animal or robot that is clearly not a human. There is a wide variety of presentations to choose from that lay outside of the Uncanny Valley. Those falling within this valley must be avoided.

If slavery of any kind is a bad idea, fully humanoid robots have through-the-roof carbon footprints, and are only in society's far-off future, what other approach might address our climate problems? How can we present this option to our young people?

NOVEL FEATURING A HUMAN/AI SYMBIOSIS

Today's young people will have to address many problems caused by Our Climate Crisis, including sea level rise, catastrophic storms, and disruption of farming. They will need all the new tools they can get for this historic task.

Young people do not read technical papers and rarely watch science-based television shows. They are much more likely to accept new information from TV series, movies, and video games. The critical information on addressing Our Climate Crisis needs to be provided in a format that young people actually use.

To explore the concept of human/AI symbioses, the author has written a novel for young adults, *Born to Storms*, (Riley, Amazon books 2020; Rukeyser, 2018) that explores this subject. The novel is specifically written to become the basis of a TV series or movie franchise.

This idea is not obvious and deserves some exploration:

Figure 2. Book Cover of Born to Storms

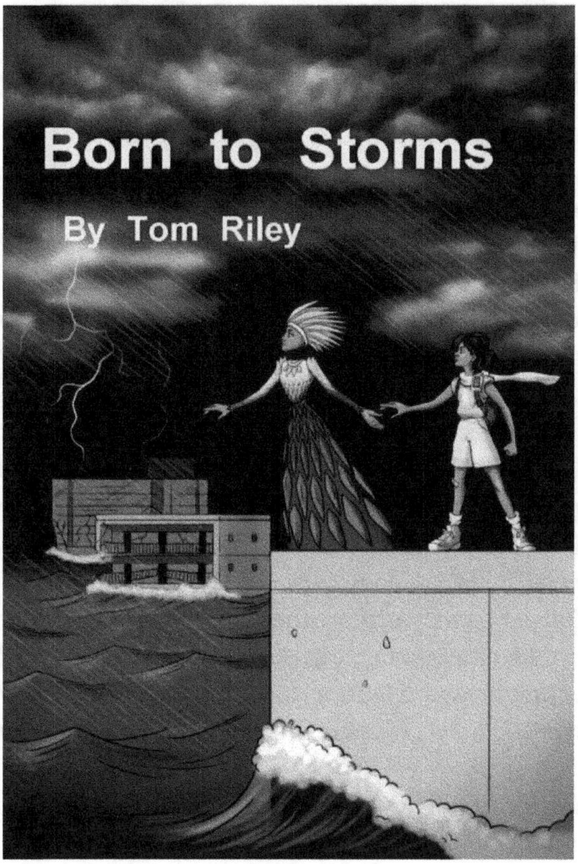

LOGLINE

Just as a technical project needs an elevator pitch, a writing project needs a logline. This is one long sentence describing the work:

Born to Storms

Set in the 2020s, a young woman, Sarah, supported by an Artificial Intelligence, JanetA, driven from her home by storms and rising seas, embarks on a life-affirming struggle to find and support many people in action on our climate crisis.

Only if this short statement has captured the reader's interest is the reader likely to continue reading the story or watching the TV series. Just like the elevator pitch, the developer must sell the idea in a few words before the project can progress.

SUMMARY OF PLOT

A one-sentence summary is hardly sufficient to provide an understanding of the work. Some deeper explanation of the plot and characters is needed.

The plot of the novel, *Born to Storms* (Riley, 2020) may be summarized:

The protagonist, Sarah, is an African-American young woman of high school age. Sarah and her AI companion, JanetA, are a symbiont pair. This interaction makes them much more powerful than either would be alone. This enhanced power then lets them participate in a great many activities to address various problems in Our Climate Crisis (Hawken, 2017; UNSDG, 2030).

The story's adventures start with a great storm that disrupts their tranquil life on a Florida beach and forces them to start their journey by requiring them to relocate. There are likely to be more and bigger storms; most years will not stand out, but a few, like 2020, will set records. The record years will come often enough to make rebuilding after rebuilding uneconomic. A great many references are available on the effects of storms and sea level rise on the habitability of locations (Horton, 2017; Englander, 2013; Jacobsen, 2019; JPL, 2020).

Sarah and JanetA's next adventure takes place in a protest march at a coal-fired power plant. In that action they explore the limitations of nonviolence. An old plant is being kept open under a temporary clause in the power agreements for years longer than promised so that it can be used as a cash cow. The protesters look at several defensive measures and how a measure intended for defense can inadvertently turn into an offensive element.

In the first summer Iron Seas adventure, they look at fertilizing the oceans with minerals to sequester carbon. The primary element is iron, which supports plankton blooms. Some of the minerals in the shells of the microscopic plants and animals then settle to the bottom for the long term. Think of the White Cliffs of Dover. Would such an effort help or hurt our oceans? Would it actually sequester carbon? They look at the level of scientific monitoring that it would take even to know if such an effort had positive or negative effects.

In Sarah and JanetA's no-till adventure, they look at what it will take to keep a small farm in business and still be a net carbon sink. Mimicking the ecology of grasslands can result in substantial sequestering of carbon in the plant matter in the soil and the retention of rainwater. This effort includes growing crops without ever plowing a field, and raising animals much closer to the wild experience of the animals' ancestors. A number of good references on this movement are available to the reader (Savory, 2013, 2016; Brown, 2018).

One of the pair's adventures is set completely on the Internet. They spend hours in a great game that teaches them how ecological systems work and what can be done when they are damaged. It is a learning experience in how the various species that make up an ecosystem support each other. The game is a study in the Serengeti Rules (Christakis, 2016). "Serengeti" means "endless plains" in the Maasai language (Attenborough, 2020). The intent of the game is to make the real grasslands and forest of the Earth again endless.

The pair next have an extended summer adventure fighting wildfires. Wildfires are yet another way that disrupted weather patterns can lead to destruction of infrastructure and the loss of human life (Maclean, 2017). In a perfect-storm fire season, like on the United States west coast in 2020, first a wet season must occur about a year before. The plant undergrowth has a banner year. Then a prolonged hot, dry season starts with the expected rains simply not arriving. High winds and dry lightning start the conflagration.

Another member of Sarah's family travels to Bangladesh to help there. The problems of a developing country hard hit by sea level rise are understood through the story. Featured are the very personal problems of sea level refugees.

In the postscript, the story jumps ahead a few years to show the pair on an expedition to the Antarctic (Fischetti, 2019; Wadhams, 2017). An AI-controlled submersible, Fishy-Fishy-Fishface14, makes an extended trip under the edge of the ice.

All in all, like all good hero's journeys, *Born to Storms* is one great adventure after another.

But is that enough? What more must the story be if it is to provide a useful tool to aid our young people in addressing Our Climate Crisis?

LOVE STORY FOR EARTH

This story may be read as a love story for Earth. Earth has been our home for about two hundred thousand years (Attenborough, 2020, pp 19) and will remain our planet of origin as long as humans and the Earth last. In the face of Our Climate Crisis, many of our young people are open to showing renewed love for our Earth (Attenborough, 2020; Wilson, 2017; Dawkins, 2004; Gould, 1989; Christakis, 2016; Pinker, 2011; Rosling, 2018).

For most of that two hundred thousand years, Homo sapiens were hunter gathers. They lived in small villages that were moved every time the local foodstuffs were depleted. Just envision coming through a pass, perhaps accompanied by a dog, and seeing spread out before you a green valley with the promise of food and good water. Your family, your village, will be fed and happy for the next season. Think of the joy and love for the environment a person would feel just looking over that valley.

Homo sapiens evolved to quickly identify and become happy over the discovery of new areas that promised their wellbeing. That instinctive reaction is the love for the Earth that society has, for the most part, lost; that is the love for the Earth that society must again find if Our Climate Crisis is to be truly addressed.

An appraisal of "Our Climate Crisis" is helpful, but what makes this problem so challenging?

IMPORTANCE OF OUR CLIMATE CRISIS

Our Climate Crisis is the defining problem of the 21st century (WMO, 2019; Hansen, 2011; Wadhams, 2017; Gore, 2006). It will generate many problems from sea level rise to a rise in the frequency of catastrophic storms. Society has left its young people with a great many messes to clean up. They will need all the new tools they can get.

A number of established climate scientists now provide ongoing sources of reliable data and commentary for the public. These include James Hanson (Hanson, 2020) and Paul Beckwith (Beckwith, 2020). Their Web pages and their supporting social platforms provide up-to-date information in a format that people can easily access.

The current COVID-19 pandemic provides a deadly example of an exponential threat. Such challenges can be addressed if confronted early, before they start doubling. If, however, society waits until the problem grows to become unmanageable, then great damage will be done.

Several of the climate crisis problems, such as sea level rise, are exponential functions. Fortunately, they have decade time constants and not just months, like COVID-19. It is important that society's young people get into action as soon as possible.

Figure 3. Projected Sea Level Rise (NASA, 2019)

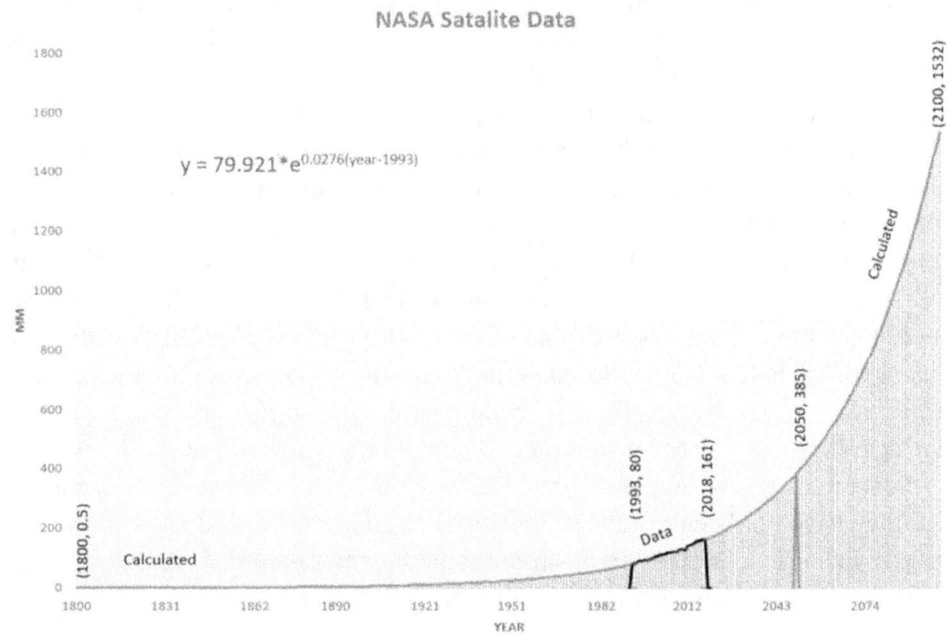

In short, COVID-19 is a hard lesson learned the hard way that society needs to take to heart.

The specific challenge that opens *Born of Storms* is not COVID-19, but sea level rise.

SEA LEVEL RISE AS EXAMPLE

Sea level rise can be taken as an example of the serious nature of the problems society's young people must face within this century. Estimates for this rise vary, but a simple exponential fitting of the satellite-measured mean sea level [See Figure 3] puts the level at about one-and -one-half meters by the year 2100 (Horton, 2017; Englander, 2013; Jacobsen, 2019; JPL, 2020).

This rise would flood out many major cities including New York and Miami. Work has already started to address this specific problem in both cities.

In New York, a sea wall is being planned for the entire lower end of Manhattan Island. This work will prevent, for at least several decades, a repeat of the damage caused by Hurricane Sandy in 2012. The work will cost several billion dollars and there is no guarantee that the protection, thus bought, will last more than a few decades.

There are also studies for a barrier across the entire mouth of the Hudson River. This type of work has been successfully done in the Netherlands, but the project would cost about one trillion dollars. Moreover, it is not clear how many decades of protection this expenditure would buy.

Miami lies on a foundation of ancient permeable coral. No barrier would provide protection as sea water easily passes through this rock. Instead, infrastructure **not** on islands of accidental high ground, (and they are only high by a meter or two), is being abandoned to the King tides that occur once or twice a month. The roads connecting the islands are being raised, at considerable expense. In addition, the local groundwater is in danger of contamination by saltwater, so fresh drinking water needs to be piped in from some distance. One resulting social problem is that low-income districts that happen to be on high ground are being bought out for more expensive housing, leaving the low-income workforce no place to live. Again, it is not clear how many decades of protection these large expenditures will buy.

So society's young people face many serious problems, but they are not responsible and they have not been presented with answers. They will need all the new tools that they can get as well as great stories about how to proceed.

Is there a story type that aligns well with this challenge?

THE HERO'S JOURNEY

Fortunately, the *Born to Storms* story type has a name: a Hero's Journey.

The idea of these classic stories was made popular by Professor Joseph Campbell with his *The Power of Myth* TV series, broadcast in 1988. (Campbell, 1947; 1988; 1991; 2008}

The classic example of this story type is Homer's *The Iliad (Homer 2015)*, written down around 800 BC. In more recent times, this story type formed the basis of such movie and TV franchises as *Star Wars* and *Star Trek.*

In the Hero's Journey, we meet a young person who is living his or her own normal life as best they can. Forces far beyond our hero's control then force the person into action. He or she then gathers a team and faces large obstacles that occur to the team as great adventures. Sometimes they win, sometimes they lose, but they have all committed their lives to a great fight.

This is the classic story that our young people need and want today. This is the story that is presented in *Born to Storms.*

Also, although replete with challenges, this is not about a failed society but one that can be saved.

NOT A DYSTOPIA

A new genre in literature, chi-fi, covering Our Climate Crisis was well established by 2013. (Irr, 2017) These stories are typically set in the near future and depict many of the great problems that society can expect.

Unfortunately, many of the novels under chi-fi that are currently available to young people are dystopias, *The End We Start From, The Big Melt,* and *American War: A Novel* (Hunter, 2017; Tillman, 2018; Akkad, 2018). Even though well written, they depict a failed and degenerate society after an ecological collapse.

These works do not provide an action plan to address the great problems. Quite the opposite, they only provide an image of loss and devastation. This is not a useful story path to present to society's young people. *Born to Storms* is not just another one of these novels. It takes a quite different approach.

Could the author just make Our Climate Crisis a villain and tell the story of its defeat?

NOT A COMIC BOOK

Our Climate Crisis cannot be made the villain in a comic book story in an attempt to reach young people. Making it a villain only promotes the myth that some evil person caused these problems and that that person can be defeated and all will then be right.

Yes, there are villains in this story. A few are the people who disparaged the Earth for short-term profit, and people who deny the severity of the problems just because they do not want to face them.

Neutralizing such people will not solve the problems of Our Climate Crisis, and presenting this as a solution will not spark our young people into action to address these problems.

What, then, can an author use to draw the interest of young people?

AI AS SHTICK

It is clear that for any theatrical endeavor to succeed financially, strong control over the look and defining key plot elements is needed. This protection has referred to by different names in different fields such as "canon", or "The Franchise". The term many technical people prefer is "Shtick".

Shtick is one of the fun words taken from Yiddish, like Schmuck. Originally stick referred specifically to the key elements of a comedy act in the Yiddish theater and then it moved to use in vaudeville and the Summer Circuit in New York State.

Now shtick has been hijacked to refer to a plot element that sets the story apart and defines all the stories in the franchise. In *Star Trek*, the shtick is the Enterprise and the transporter. In *Star Wars*, it is the light saber and the Force.

Born to Storms features a symbiotic relationship between a Human and an Artificial Intelligence as its shtick. Thus symbiosis is the shtick franchise and must appear in any books, movies, TV series, or games in the series. This relationship is radically different from the master/slave relationship currently envisioned for most AIs and therefore can define the franchise.

The possibility of an AI revolt has been often used as a major shtick in science fiction; just consider *Terminator* to *Blade Runner*. As discussed above, human slavery has been tried many times, and these efforts have often produced disastrous results. Why should AI slavery prove any different and why would the resulting action plot make for such common stories?

But what are the alternatives?

The Born *to Storms* shtick features a person and AI pair that are matched at a young age. They are then trained as a symbiotic pair for life. There are not a large number of such pairs around, but there are enough that most people recognize a human-AI pair quickly and accept them.

The AI does not have a robotic body but normally appears as an image on a cell phone that the human wears on the shoulder. The AI can, however, jump to any monitor that the human is using at the time.

However, the AI in our story, JanetA, is not truly conscious. She would fail the common tests as she often trips up in ways a five-year-old human could avoid.

The AI is not given a robotic body because that exceeds the technology likely to be available in the near future, and such a body would have a very large carbon footprint.

The key here is that the symbiont pair of human and AI is much more powerful than either one would be alone. Together they can take on the great problems of Our Climate Crisis.

Overall, *Born to Storms* attraction depends on the words of Arthur C. Clarke for its premise:

"Any sufficiently advanced technology is indistinguishable from magic."

The magic of the promise of technologies that may soon be within the reach of the readers is the magic that fires this story.

How close is society's technology to the AIs in this story? Could a symbiotic approach be developed in time with a low carbon footprint?

TECHNICAL REQUIREMENTS

The technologies called for in the novel were chosen because they are likely to be developed in the near future and have a low carbon footprint. The key technologies used are:

1. Better Batteries – New battery technologies with four times better capabilities as appreciated by the user. This advance will make electric cars more desirable and make personal electronics very convenient to use.
2. Deep Learning Chips – Dedicated chip sets with custom algorithms. This will reduce the power requirements and physical size of Deep Learning equipment, making AIs even more widely available.
3. High Speed Internet – The 5G network should suffice. The proposed human/AI symbiont must communicate between its various parts quickly and with wide bandwidth.

In short, the technologies needed for the human/AI symbiont in the story are either already available or currently under advanced development.

That said, many technologies have developed problems and controversies that greatly limited their value.

CONTROVERSIES

Any great story features controversies to add interest to the plot. The idea of a human/AI symbiosis generates important ethical concerns:

1. Experimenting on Children -- Changing kids' lives as part of an experiment is a major ethical concern that requires considerable planning and close monitoring. Establishing a true symbiont will require starting when the human is young and training for years. In one sense, society now provides children with years of schooling, and experimental teaching procedures are tried from time to time. Clearly, the effort will have to be limited at first and independently monitored. An exit plan will also be critical.
2. Kids control AIs – Depending on kids to control AIs when many adults are clearly afraid that the AIs will run amok. Is society placing a burden on its children that it does not wish to address?
3. Size of test -- How many human/AI symbionts would be needed for a definitive test? In the story, the number is over 8,000.

4. Fair representation -- Will 8,000 test pairs represent the population fairly? In the story, the price is described as more than a car but less than a house. Will rich people buy their children into the program? Will funding agencies finance enough poor people to prevent the further development of an uber-class?
5. Unforeseen problems – What problems of Our Climate Crisis could a human/AI symbiont take on better than if they were in a master/slave arrangement?
6. Better AIs – The human/AI symbiont test will certainly result in improvements in AI hardware and training. Will the general population trust the program to provide protection from an AI revolt while addressing major societal problems?
7. The End of Homo sapiens – At some point, a human/AI symbiont should be classified as another species. Is society rushing toward this point?

Any approach society takes to address the problems of the 21st century is bound to generate some controversy. One current question, then, is if the human/AI symbiont idea is worth pursuing.

Has society learned any lessons from its hard COVID-19 experience that will help resolve these promises and controversies?

LESSONS LEARNED FROM COVID-19

Our climate crisis will force many changes to society. Sea level rise alone will require the relocation of many large cities from New York to Calcutta worldwide. The same goes for changes in the local weather that will make growing food difficult. If a person is a farmer, or fisherman, then knowledge of the seasonal weather is critical to their livelihood and thus critical to having enough food for the population as a whole.

As the spread of COVID-19 virus was an exponential function, the pandemic has provided a practice run for addressing great problems, if society understands this challenging experience as such. As soon as the science of COVID-19 was understood, many countries went into voluntary lockdown and the wearing of masks was adopted widely. Some countries adopted strong measures and other countries delayed or avoided adopting these practices altogether.

The requirement clearly demonstrated the characteristic of an exponential problem. People can adapt and adapt quickly when a threat is understood. Society must, however, identify the problem early and take action while action is still effective. This requires leadership, and a good story can only help.

When the story of COVID-19 is written, what form will it take and what will be the lasting effects?

POSSIBLE AI EFFECTS ON SOCIETY

Since the beginnings of the Industrial Revolution, around 1800, human societies have looked to improvements in technologies to address major problems. From the steam engine to the Internet, this vision marched pushed on. Yes, sometimes a new technology did more harm than good, but the powerful ones pushed through to envelop most societies, one way or another.

This brings us to the human/AI symbiosis proposed in the novel *Born to Storms*. Could this be a significant element in this century of great change? The required technologies are either here or close at hand. The result to society would be quite different from the result expected if Master/Slave is unchallenged.

Can society take the necessary actions to avert the great problems of Our Climate Crisis, or will action have to wait until there is a clear and desperate emergency?

CONCLUSION

The Master/Slave relationship for humans and their machines is **not** satisfactory, both in its name and in the technical process. As the capabilities of smart machines grow, a new interactive relationship is needed. Symbiosis could be that new arrangement, but a considerable development effort will be needed.

Our Climate Crisis will force great changes to human societies by the end of this century. Society needs all the new tools it can get. Artificial Intelligence is clearly one of these tools, but how shall it be used? Who will reap the benefits? What contribution can it make to address the great problems of Our Climate Crisis?

These questions will only be answered through time, but to be effective, action must start now. The real possibilities for addressing them must be made clear and compelling to society's young people.

REFERENCES

Akkad, O. E. (2018). *American War: A Novel*. Vintage.

Attenborough, D. (2020). *A Life on Our Planet*. Grand Central Publishing.

Barber, J. (2018). *Imagining Sustainable Futures in Popular Culture: Moving Beyond Dystopia and Techno-fantasy Narratives*. IAMCR 2018 Conference, Eugene, OR.

BeckwithP. (2020). *Home Page*. https://paulbeckwith.net/

Brown, G. (2018). *Dirt to Soil, One family's journey into regenerative agriculture*. Chelsea Green Publishing.

Burkov, A. (2019). *The Hundred-Page Machine Learning Book*. Andriy Burkov.

Campbell, J. (1991). *The Power of Myth, with Bill Moyers*. Anchor Books, Random House.

Campbell, J. (2008). *The Hero with a Thousand Faces*. New World Library. (Original work published 1947)

Campbell, J., & Moyers, B. (1988). *The Power of Myth [DVD]*. Kino Lorber.

Christakis, C., & Sean, B. (2016). The Serengeti Rules, The quest to discover how life works and why it matters. Princeton University Press.

CNN. (2003). *'Master' and 'slave' computer labels unacceptable, officials say*. CNN.

Dawkins, R. (2004). *The Ancestor's Tale, The dawn of evolution*. Houghton Mifflin.

Englander, J. (2013). *High Tide on Main Street, Rising sea level and the coming coastal crisis*. The Science Bookshelf.

Fischetti, M. (2019, August). A New Reality, Climate Change is dramatically altering life at the top of the world. *Scientific American*, *321*(2), 37.

Gore, A. (2006). *An Inconvenient Truth: The planetary emergency of global warming and what we can do about it*. New York: Viking.

Gould, S. J. (1989). *Wonderful Life, The Burgess Shale and the Nature of History*. W. W. Norton.

Hansen, J. (2011). *Storms of My Grandchildren, The truth about the coming climate catastrophe and our last chance to save humanity*. Bloomsbury Publishing.

Hansen, J. (2020). *Home Page*. Retrieved on October 2020, https://csas.earth.columbia.edu/

Hawken, P. (2017). *Drawdown, The most comprehensive plan ever proposed to reverse global warming*. Penguin Books.

Homer. (2015). The Iliad (C. Alexander, Trans.). CCO HarperCollins.

Horton, T. (2017). *High Tide in Dorchester, A Bay Journal Documentary*. Retrieved from https://hightidedorchester.org/

Hunter, M. (2017). *The End We Start From*. Grove Press.

Irr, C. (2017). *Climate Fiction in English.* Retrieved from http://www.academica.edu/35288142/Climate_Fiction_In_English

Jacobsen, R. (2019). Beyond Seawalls, Fortified wetlands can protect shorelines better than hard structures. *Scientific American, 32054*, 32.

Jet Propulsion Laboratory, Physical Oceanographic Distributed Active Archive Center. (2020). *Seal Level.* Author.

Kellener, J. D. (2019). *Deep Learning*. MIT Press. doi:10.7551/mitpress/11171.001.0001

Kidder, T. (1981). *The Soul of a New Machine*. Little, Brown and Company.

Maclean, N. (2017). *Young Men and Fire*. University of Chicago Press. doi:10.7208/chicago/9780226450490.001.0001

Riley, T. (2020a). *Born to Storms*. Amazon.com.

Riley, T. (2020b). *Born to Storms Web Page*. Retrieved from website at https://bigmoondig.com/Stories/BMDStoriesBtS.html

Riley, T. (2019). *Sea Level Rise Extrapolated from current NASA readings*. NASA. ftp://podaac.jpl.nasa.gov/allData/merged_alt/L2/TP_J1_OSTM/global_mean_sea_level/GMSL_TPJAOS_4.2_199209_201803.txt

Ripple, W. J., & Wolf, C. (2019). *World Scientists' Warning of a Climate Emergency*. Retrieved from https://academic.oup.com/bioscience/advance-article/doi/10.1093/biosci/biz088/5610806

Rukeyser, M. (2018). *Science Fiction as the Mythology of the Future. The universe is made of stories, not of atoms*. https://www.centerforfutureconsciousness.com/pdf_files/Readings/ScienceFictionLongArticle.pdf

Russel, S. J., & Norvig, P. (2015). *Artificial Intelligence, A modern Approach, Third Addition*. Person Education.

Savory, A. (2013, February). *How to green the world's deserts and reverse climate change*. Retrieved from TED video at https://www.ted.com/talks/allan_savory_how_to_green_the_world_s_deserts_and_reverse_climate_change?language=en

Savory, A., & Butterfield, J. (2016). *Holistic Management, A commonsense revolution to restore our environment*. Island Press.

Seattle PI. (2004). 'Master/slave' named most politically incorrect term. *Seattle PI.*

Theobald, O. (2017). *Machine Learning for Absolute Beginners*. Second Addition. Theobald.

Tillman, N. (2018). *The Big Melt.* South Branch Press.

United Nations Sustainable Development Goals. (n.d.). *Transforming our World: The 2030 Agenda for Sustainable Development.* United Nations. Available at https://www.un.org/sustainabledevelopment/

Wadhams, P. (2017). *A Farewell to Ice, A report from the Arctic.* Oxford Press.

Wagner, G., & Weitzman, M. L. (2015). *Climate Shock, The economic consequences of a hotter planet.* Princeton University Press. doi:10.1515/9781400865475

Wilson, E. O. (2012). *The Social Conquest of Earth.* Liveright Publishing.

Wilson, E. O. (2017). *Half Earth. Our Planet's Fight for Life.* Liveright Publishing.

Wired. (2020). Tech Confronts Its Use of the Labels 'Master' and 'Slave'. *Wired.*

World Meteorological Organization. (2019). *The Global Climate in 2015-2019.* Retrieved from https://library.wmo.int/doc_num.php?explnum_id=9936

Chapter 8
Artificial Intelligence, the Risk to Do Something Against the Man:
Innovation Must Be Governed to Avoid Disgregation

Anna Verrini
Indipendent Researcher, Italy

ABSTRACT

A long time ago we started to speak about artificial intelligence. It was like a dream for many years, and now, with the tremendous computing capability and data availability, we are living in a kind of reality where we think everything is possible because we can train machines to do more or less everything man is able to do both with his hands and with his brain. But we must keep in mind that we can train machines to do what we do, in the way we do, but it is quite difficult, for example, to train machines, and not only, to take decision in the way and with the quickness usually necessary. And what about ethics? What kind of future can we design? This chapter explores the risks of artificial intelligence.

INTRODUCTION

Speaking about Artificial Intelligence is quite mandatory to start with Marvin Minsky, one of the pioneer of Artificial Intelligence.

DOI: 10.4018/978-1-7998-7126-2.ch008

Figure 1. Marvin Minsky open

Minsky, American Mathematician and Cognitive Scientist, was born in New York in 1927 and died in Boston in 2016. He was largely concerned with research on Artificial Intelligence and was co-founder of the Massachusetts Institute of Technology's AI laboratory. Anyway, it is not the case to proceed with the Biography of Marvin Minsky, very well known to the majority of the people with some interest in AI, Computer Science, Innovation. He has received the ACM Turing Award, the MIT Killian Award, the Japan Prize, the IJCAI Research Excellence Award, the Rank Prize and the Robert Wood Prize for Optoelectronics, and the Benjamin Franklin Medal.

Minsky was adviser on Stanley Kubrick's movie *2001: A Space Odyssey* and he was also mentioned explicitly in Arthur C. Clarke's derivative novel (Clarke, 1968) with the same title. He argued that "somewhere down the line, some computers will become more intelligent than most people," but that it was very hard to predict how fast progress would be. He cautioned that an artificial superintelligence designed to solve an innocuous mathematical problem might decide to assume control of Earth's resources to build supercomputers to help achieve its goal, but believed that such negative scenarios are "hard to take seriously" because he felt confident that AI would go through a lot of testing before being deployed (Jerusalem Post, May 13, 2014).

From the very beginning science and science fiction started to move forward together in a way where science fiction seems to anticipate what will happen sometime in the future without knowing exactly when. Many times the stories of the science fiction become real life with some modification, of course, but remaining very close to what has been created in the fiction. Everyone must be convinced that what would be better to do is that man keep in their hands the ability to design and write their future (Minsky, 1967), (Minsky, (Minsky, 1986), (Minsky, 2006).

And this is what anybody would believe: the worst scenarios cannot become reality because anybody is convinced, and hopes, that the human mind will keep the governance in the innovation and in the evolution (Stephen Hawking, The Guardian 2014-dec-2), (Elon Musk, Vox 2018-nov-2).

THE COMPUTER FROM THE BEGINNG

From the very beginning computer had the capability to strike the imagination of people and to open futuristic views. Computer is usually the "star" in a science fiction novel and it is demonstrated that is a character that can draw a lot of readers. Also in the everyday life computers and people working with a computer can fascinate a lot of persons.

From the beginning of the human history, more or less 2.100 years before Christ in China and then for ancient Greeks and Romans, abacus was a help for the man to make calculation. A first trial to design and build a general purpose computer was the Analytical Engine by Charles Babbage.

Middle 1800 Ada Lovelace, the daughter of Lord Byron, added some Notes to an article of the Italian engineer Luigi Menabrea containing what many consider to be the first computer program—that is, an algorithm designed to be carried out by a machine, which included a way to calculate Bernoulli numbers. That was widely considered to be the first complete computer program and Ada Lovelace is still now very famous being described as the first computer programmer. Turing started from the Menabrea studies with the integration of the notes of Ada Lovelace to build the first computer after the second worldwide war. A computer program language was called ADA in honor of Ada Lovelace (Wikipedia Biography).

The very fast technical evolution brought to tools more and more powerful until today when we are in the middle of a disrupting innovation growing up with an exponential trend and involving any field of the human being from the work to the family life to the social life. But it is important to say that, anyway, it is a tool designed by man and then it would be mandatory to keep under control of man.

At the moment, the explosive evolution of ICT is going on both for computational power, for capability to organize and manage a tremendous amount of data to create information and for communication speed. This gives the possibility to release new applications that can give us an essential support in a moment like the one we are

Figure 2. Charles Babbage machine open

Figure 3. Ada Lovelace open

living now for an invasive innovation in every field of the human being. We are speaking about Industry 4.0, Smart Working, Health 4.0 and so on.

Industry is evolving very quickly through the benefit coming from the use of innovative tools capable to create improvement in any area:

- The production process.
- The maintenance of production facilities.
- The sales activities.
- The HR management
- Logistics
- And many others

just to give some examples where innovation is specifically important in the production and service companies.

INNOVATION TODAY

The improvement created through innovation in every field is completely changing the life of the man for, more or less, any type of work. For tiring works man and woman can be replaced by machine that can do the same job faster and with more reliability and security. More than this, specifically for repetitive jobs, there is no risk to do something wrong due to a lowering of attention like many times happens for repetitive jobs. We can do thousands of these examples and we can say that life is really bettering due to innovation available everywhere.

We have the possibility to get a better time balance between family, work and personal life then spending more time with our relatives and our personal interests. Probably this is one of the best results anyone can get from the new way to work. Also in the day by day life we have a lot of services created to solve problems very easily without losing time, like possibility to use any financial service through the online bank, to buy any kind of tickets from the air to train/bus, theatre ticket, the possibility to work some days from home, to organize and do meetings with colleagues, suppliers, customers not being obliged to move to different sites. And it would be possible to go on for pages to mention any kind of solutions helping anyone both in the personal life and in the work life. But now it is possible to assist to a very important phenomenon, that is the way the majority of industries, school, university solve the problem to keep at a security distance each one from everyone working from remote in a moment when this is probably the only way to keep COVID epidemic under control. Then, connecting from remote, we can work, study, contact colleagues, friends, it is possible to live, let's say, a normal life. But, also in this case, we must start to govern the situation because it is impossible to believe we can go on forever without contacts between persons. It is necessary to start to design the future with a mix of remote and direct way to speak, work, study among persons.

Innovation available now give a lot of capabilities in many fields to do new things, or to renew the way to do getting better results. It is very important to give some examples.

Let's refer to 3D Printers. There is something very interesting to emphasize, that is the use of the 3D printer in the cultural heritage environment to restore, to preserve artistic works, to organize tactile path for blind persons in the Museums.

Other interesting but completely different applications could be organized using 3DPrinters for the management of the spare parts. Middle '80 very interesting application started using the first networks (not yet internet, but timesharing network with a worldwide communication coverage) to move orders and information for goods and merchandise from different remote sites of large and distributed companies to the headquarter. Just to give an example, it is interesting to mention the case of car producing companies where it was possible to connect the headquarter with the car dealers to receive and process more quickly the order for cars and procurement of spare parts when and where necessary. Going on innovation very quickly now it would be possible not only to receive procurement request for spare parts but also to produce the parts, where necessary, using 3D Printers. And this could be a disruptive solution not only for the car producing companies but also for companies producing, let's say, refrigerators, washing machines and any kind of machines where it is necessary to guarantee the maintenance to the final customer doesn't matter where he is. The advantages could be enormous both for the business logistics of companies and help to protect the environment for the consequence on pollution.

Using 3DPrinter it is possible also to produce human organs and this is excellent as alternative to replace one in a human body without manage a waiting list to get organs through the death of persons for patients who are living in a very critical situation. It is quite difficult to accept that it is necessary the death of a person to save the life of another. Now the question is: if something goes wrong in a transplant operation, where is the problem? The materials used to produce the organ, the computer programs for the 3D printer, something wrong with the surgical team? The problem can be everywhere: only detailed controls can give us the possibility to understand, with an acceptable probability, where is the problems.

There are also other general kind of benefits we can get now due to the fact that, in days when it is very critical the worldwide situation because of the climate changes, it is possible to start to reduce air pollution with less persons going around by car, airplane, train and so on, and to do more controls both for waters, cultivable land, pasture land. For sure there is a lot of thing we can do now to reduce consumption of energy, thermal pollution and control many other sources of pollution.

We will be able to help more and more women to make domestic works giving them possibility to do something more interesting and useful for the society: now domotics, that is home automation, is a very well developed area, from the availability of small robot cleaning floors in the houses to the remote control for washing machines, heating systems, refrigerators, just to mention some of the features and then the benefits.

And let us think about what more we can do in the healthcare area. Many ways are available to improve the quality of the health of the persons both for the improvement of the organization and for typical medical application. Medical wearable equipment gives the possibility to realize checks on critical parameters that can be continuously sent to a center where data and information are immediately elaborated sending back alerts or, when possible, parameters for corrective actions. Typical solutions refer the control for blood pressure, glycaemia, heartbeat and many other health problems. These applications give possibility to old people to lead a more independent life and younger persons, still working, to lead a more normal life. Usually centralized services are organized such a way to get as soon as possible also the assistance of specialized personnel, when necessary.

More than specific applications that it is possible to find in this area, what it is very important is the support that big data applications can give to the doctors for specialized diagnosis. We have now the possibility to collect, store, manage and process a tremendous amount of data from everywhere in the world and obtain correct information to support specialists in their work virtually in any field. Speaking about healthcare, a very important application is related to the possibility to collect information concerning any kind of disease, from the diagnosis to the therapy, to the results of the different therapies.

This means to give doctors a very important tool to make the diagnosis, specifically for rare diseases or illnesses particularly complex to diagnose. Of course, it is not possible just to accept the result given from the application, but it is necessary to consider only as suggestion with a known level of probability because it is absolutely necessary that the doctor matches information from the system with the analysis of the person who is in front of him.

Not only in the healthcare area, but also in many other fields AI applications are considered from many persons a kind of "statistical applications" giving the level of probability that something is true or can happen. What does it mean? In the healthcare application, for example, a diagnosis cannot be accepted as is coming from the system, but only as a suggestion that could be a specific disease "with a probability, let say, of 98%": only the doctor can confirm because he has the patient in front of him and he can judge through elements that can be sensed only personally analyzing a man under many respects using what we call the soft skills.

A comment now comes up naturally: the new way of working should be a work team including all the specific professionals from the production to the logistic, from the human resources to the sales, to the doctors and computers experts.

THE PLUS AND MINUS OF INNOVATION

Until now we have seen that innovation is completely changing and improving the human life under any point of view. As it happens every time when living an industrial revolution, we start immediately to be afraid that we can lose or see a reduction of our jobs due to new machines and new solutions. Between '800 and '900 the transport system has been completely revolutionized. In the beginning it seemed that a lot of persons could lose their jobs, but it was true exactly the opposite because jobs changed but did not disappear.

The Railways, airplanes, cars and so on gave the possibility to connect every part of the world very quickly moving persons and goods. Jobs changed but the number of persons involved increased in a very significant way. Now we have tremendous problems with the climate changes we see from years due to pollution produced from a number of causes coming from transport systems, industry and so on. After a long delay we are trying now to contrast what is happening and hope to be on time.

Now, with the digital revolution involving every field of the human life, we have a lot of jobs that can be done without the human participation. That's good because people can make something clever and more interesting than tedious or tiring jobs. But we must be sure that everything has been anticipated and planned in advance, otherwise we can get problems that we cannot solve because could be impossible to change in time what have been automatized.

Going on, what it is very important is the capability to upgrade continuously our professional profile together with the disrupting speed of the digital evolution. It is also very important to be capable to organize better our life in relation to the family and social life.

Also with the social network systems it is mandatory to be very careful because these are powerful tools that can create enormous damages if used with no governance, from the possibility to condition millions of people through fake news and ideas, to the possibility to give young people the illusion to have millions of friends without being sure also of the correct name of the friends. Worst case than this, what is very critical is the possibility to drive children and persons with a delicate psychological balance to make something very dangerous, with no way to go back. It is sure that we cannot eliminate the use of social network because it is very important to have this kind of solution for example in case of emergency but we must be very careful with the abuse of this kind of tool.

INNOVATION CAN ALSO DISTROY

It is not just a joke. It is something that can happen. Beginning '900 in a small area in Liguria, region in the north of Italy, there was a chemical plant, SIPE (Italian Company Explosive Products) that had an "explosive" growth with the first worldwide war. There was a specific product made for Russia that was stopped with the Russian Revolution and when hostilities ended. Some persons were capable to take the decision to change production and go on to use machineries, materials and guarantee work to the persons. With a lower degree of nitration, nitrocellulose together with camphor as plasticizer, from explosive becomes film. It was the beginning of film production and with the fantasy of some persons, the capability to decide and the support of a French company, it was possible to start a long success story going over until when pictures became digital. In the beginning other production coexisted with film, then the company had financial problems solved by a person sent to close the company. He was capable to balance the books and start with a new phase until when the company entered the IFI area, the Financial Headquarter of Agnelli family. The Headquarter moved to Milano and with the original group of young technicians

Figure 4. Ferrania Manufacturing Plant – Ferrania Museum

Figure 5. Women grow wheat in the manufacturing plant fields – Ferrania Museum

and a very strong research group, along about 30 years had the capability to grow up becoming one of the 5 companies in the words with the technology to produce film. After a strong collaboration with Cinecittà, the Ferrania product has been use to produce the best Italian films in the years immediately following the end of the second worldwide war. Workpeople grew up to 3000 persons and was always a community strictly related and supported by the general management through a CSR policy until when IFI decided to sell to the American Company 3M. Few persons of the old management remained with the new owner and the management control was taken by the American Company. Step by step production was modified, personnel reduced until when bankruptcy has been declared beginning 2000'. Nobody had the intuition and let say, the fantasy, capable to save the activity (Ferrania Museum).

CONCLUSION

And now let us make very quickly some considerations about the Artificial Intelligence that it is possible to see under two different points of view, one site as application of Learning Machines and on the other site as "emulation" of the human mind. In the first view this means to analyze in a deep detail a process and reproduce to be automatically executed by machines with the possibility to upgrade directly the knowledge of the machine itself while working. In the second view there is, in my interpretation, a very big question: why should we reproduce the human mind in all its characteristics? We must remember the sentence of Minsky mentioned at the beginning of this chapter and the opinion of many others like Stephen Hawking who alerted, on 2014, about the dangers of AI considering it a risk for the humanity: "*Artificial intelligence could spell the end of human race*" (The Guardian, Dec. 2014). Not only Hawking made this kind of assertion but also many other, and just to mention one, Elon Musk, speaking at MIT in 2014, called *AI humanity's "biggest existential threat" and compared it to "summoning the demon"*. He reiterated this opinion also during an interview when he said: *"I do think we need to be very careful about the advancement of AI."*

What is very difficult is to predict whether and when this will happen. For every innovation, in any field, after a while we discover that innovation itself gives a lot of benefits but also can create problems, risks, damages as a consequence also in many other areas, not only in the original itself.

A way to reduce risks for humanity probably is to start to introduce at design level ethics principles and invest in a deep activity of research on the possible consequences of innovation projects. We must start to apply always and everywhere the ethics principles and keep them updated, following the disrupting evolution of the information technology, as it is done at the ACM (Association for Computing Machinery), that created and is keeping updated the ethics code from about 40 years. We must keep in our hands the governance of innovation and not "to fall in love" for anything is new and projecting ourselves in the future (**Communications of the ACM, December 2016, Vol. 59 No. 12,** www.acm.org/code-of-ethics, https://ec.europa.eu/info/sites/info/files/commission-white-paper-artificial-intelligence-feb2020_en.pdf).

At the end we must take into account the help from the science fiction. Let us remember the three Robot laws from Isaac Asimov:

First Law

A robot may not injure a human being or, through inaction, allow a human being to come to harm.

Second Law

A robot must obey the orders given it by human beings except where such orders would conflict with the First Law.

Third Law

A robot must protect its own existence as long as such protection does not conflict with the First or Second Laws.

Is this enough? Probably not. Probably it is important to learn from science fiction what will happens in the future, also if we don't know when. If now we read robot stories created by Asimov, and others, a long time it is possible to find that what has been written 30/40 years ago, now happens in the real life.

A machine must be correctly designed and properly used to be sure that does not create any kind of risk. Very often science fiction has demonstrated to be an anticipation of the future that many times becomes reality. But the science fiction authors many times are scientists, so we must believe what they are saying (Filippazzi et al, 2018).

REFERENCES

Clarke, A. C. (2001). *A Space Odyssey. Hutchinson.* UK New American Library.

For artificial intelligence pioneer Marvin Minsky, computers have soul. (2014, May 13). *Jerusalem Post.*

Maggio, F. F. (2018). *Mondo Digitale: Il computer tra realtà e fantasia (Computer between reality and phantasy).* Associazione Italiana Informatica e Calcolo Automatico.

(2016, December). Making a positive impact: Updating the ACM Code of Ethics. *Communications of the ACM*, *59*(12).

Minsky, M. (1986). *The Society of Mind.* Simon and Schuster.

Minsky, M. (2006). *The Emotion Machine.* Simon & Schuster.

Minsky, M. L. (1967). *Computation: Finite and Infinite Machines*. Prentice-Hall.

Novembre, F. F. (2017). *Mondo Digitale: Intervista a Eugene Goostman – Divagazioni sulla Intelligenza Artificiale (Interview to Eugene Goostman – Digressions on Atificial Intelligence)*. AICA Associazione Italiana Informatica e Calcolo Automatico.

On Artificial Intelligence – A European approach to excellence and trust. (n.d.). White paper. https://ec.europa.eu/info/sites/info/files/commission-white-paper-artificial-intelligence-feb2020_en.pdf

Chapter 9
Knowledge Obsolescence and the Future of Work:
Relevance of Knowledge and Impact to Jobs

Bobby Varanasi
Matryzel Consulting Sdn Bhd, USA

ABSTRACT

The chapter delves into a range of influencing factors that are governing individual and corporate behavior, driven both by changing human circumstances—economic, social, and environmental—as well as rapid changes to organizational cultures and endeavors—models, markets, governance, customers, competition. A veritable confluence of these factors is impinging into workplaces in a never-before-seen manner, particularly in its inherent complexity and constant change. Jobs are being redefined, created, and eliminated at the same time, putting significant pressure on individuals seeking to pursue careers. Knowledge acquired over a certain period is becoming quickly obsolete, resulting in people having to shift gears quickly. Most fail, with consequences impacting both social structures and organizational cultures. Individual behavior is significantly deteriorating toward colloquialism driven by a sense of victimization. How do we address all these challenges and stay on top of the future? This chapter's aim is to distill the answers to this question.

DOI: 10.4018/978-1-7998-7126-2.ch009

INTRODUCTION

The world is awash with new technologies, being launched by nimble startups and large corporations, on the back of a range of discussions around disruptions, new value creation with consumption models and the ever-present promise of bridging physical and virtual worlds. Alongside these exciting developments, a few questions arise, spanning the economic and socio-political spectrum. While the most notable deliberations are around changing nature of jobs because of automation and artificial intelligence, more mundane conversations demanding that corporations change their entire approach toward managing millennials has constantly occupied our current thinking. Alongside, we have been witness to concerns raised around increasing inequality, stagnation of (wage-driven) incomes worldwide, and replacement of jobs by machines (increasing the push-back on adoption of automation technologies). Meanwhile governments are beginning to come around to considering alternate solutions like universal basic incomes, restructuring public distribution mechanisms and creating new avenues for revenue generation.

Interestingly, many do argue that automation has always been there. Why then is it perceived to be different this time around? The argument is that in the past we had deployed machines to add muscle to human endeavor. In doing so, such mechanization automated routine tasks, which only complimented humans. However now we are not just leveraging intelligent machines, but they seem to be doing completely different and new tasks that didn't exist before. This fact doesn't seem to sit well with people who are beginning to get concerned about what would be left for humans to do when all this reaches a point of no return. The debates around future of work are about this journey.

Doomsday predictions have begun to spring up as it relates to the types of jobs that will be eliminated before the end of this decade. They may seem harsh and disturbing. Yet, somehow the inevitability of it has not yet dawned on people. It is therefore pertinent to take a step back and understand the context in more certain terms. This paper, and the following three papers will center round the facets influencing workplaces, jobs, tasks etc. in a concrete manner. The goal is not to predict the future. Instead, my attempt here is to establish a greater understanding such that we are able to collectively ready ourselves for the future.

WORKPLACE AMBIGUITIES

Let's set the context first. Talent is changing. Education is abundant. And technologies are creating new opportunities at the learning and deployment ends of the spectrum in a more pervasive manner. Structural changes in the workplace can be painful, but ultimately progressive. Various new techniques driven by machine learning, deep learning, and AI are opening new vistas to gaining business insights. Such tools are also being utilized – particularly in medical sciences - to resolve some thorny issues.

Deployment of such tools in businesses is resulting in greater efficiencies, but increasingly diluting the role of people as a factor of production. While this has a direct impact on "employed workforce", organizations have begun to deploy off balance-sheet talent through the adoption of an open talent continuum. These shifts are forcing us to reconsider the role of individuals, organizations, hierarchies, accountability workflows, labor laws et al. Re-imagining the workplace is possible. However questions remain on the nature of evolution, unavoidable disruptions, resistance to change, and dealing with lost opportunities. We believe this question is best addressed not by looking at the impacted, but by the forces shaping such changes – changing customer behavior, demographics, globalized marketplaces, virtual platforms and the interactions that take place amongst these forces.

We believe there are three distinct forces influencing the world we live in today. These forces are shaping the future of workplace and consequently, human behavior within contexts of society and economies. We term these forces as the Briquettes (Varanasi, 2019).

Briquette 1: Virreal World

The past decade and a half has been witness to the convergence of real and virtual worlds, thanks to the vigorous spread of the internet globally. More than half of humanity has access to the internet and is therefore able to navigate both the worlds almost seamlessly. Businesses are transacting virtually while customers consume physically. Self-learning technologies are able to predict consumer behavior far more accurately than previous management models have ever come close to. Supply chains are beginning to look more like homogeneous networks than the distributed, and severely fragmented ones they were designed to be. While there are inherent issues surrounding network effects, cybersecurity, information pandemics, and the rise of digital nomads, these are considered new world problems.

Another half of humanity seems impervious to all these changes, and remains firmly entrenched in an old world where influences from the virtual world are yet to entrench themselves in a beneficial manner. They are however unfortunate victims. To the extent that smartphones have become ubiquitous across most people worldwide, the inherent assumption that they are therefore connected to the virtual world is a fallacy at best, and irresponsible at worst. Distribution and governance systems are yet to be redesigned in a manner that the Virreal world is able to spread its benefits to all, and not just the connected ones.

Briquette 2: Workplosion

Changes in the workplace owing to adoption of global communications networks is obvious. It is a foregone conclusion that any organization today is awash with a range of technologies and devices, tools and infrastructure to help conduct business in a modern manner. Interestingly however, discussions on workplace modernizations almost always remain limited to organizational cultures, security policies, engagement of millennials, and leveraging virtual teams for production endeavors, within the context of quick obsolescence with knowledge, shrinking customer loyalty, and challenges with constant automation.

There is neither clarity nor emphasis on more fundamental issues imploding within workplaces. Trends like increased sense of victimization, digital narcissism, contextual deficit, low attention spans, high sense of entitlement, and instant gratification are slowly yet inevitably chipping away at organizational foundations in more ways than one is able to clearly put a handle on. These seemingly detrimental trends are more often than not discarded as nothing more than an inability or unwillingness to change. The deeper factors driving such behavior are neither being examined, nor understood well enough. At another end of the spectrum, moral conflicts surrounding automation and humans is yet to hit mainstream conversations. All these implosions within the workplace will need to be understood, examined and addressed proactively if we are to shape a positive future to workplaces.

Briquette 3: Borderless Boundaries

Globalization is here to stay. We have heard of this for over two decades now. Proponents have always argued for its increased value, and pursued it with vigor not seen outside war zones. Interestingly however, all such endeavors were more or less pigeonholed around management models that talked about scale, replication, standardization and virtualization. Such efforts may have worked, resulting in more complex networks of organizational hierarchies and reporting structures, alongside reduced degrees of separation.

An often-ignored reality with globalization has been the incremental creeping in of boundaries, manifested predominantly in disappointments amongst customer groups where localization wasn't prevalent, increased lack of trust amongst collaborating individuals and teams within organizations, and a sense of helplessness at the end of it all. The phrase often used seems to be that the "an aggressively pursued multipolar world order has resulted in depravity more than prosperity". While I shall refrain from arguing for or against it, I shall submit that a host of boundaries have emanated from the very nature of our endeavors at globalizing, rationalizing and creating seamlessness. Meanwhile, nation states are beginning to push back on global trade strictures – the very backbone to globalization – owing to issues surrounding mutuality and equality with sharing garnered benefits. Borderless boundaries are a reality today and need to be addressed urgently if we are to make positive strides toward the future of workplaces.

GAZING FORWARD

It becomes pertinent to envision the multi-nodal nature of various contributing factors to these three briquettes, particularly their direct and consequential impacts on one another, the interconnects, network effects and consequences to workplaces, as reflected in the infographic below.

Each of the briquettes shall need to be delved into in greater detail to understand interconnects, network effects and the consequences on workplaces. We shall be dedicating the remainder of this compendium to distilling these and bring to the fore a range of issues and opportunities that would enable us to collectively reimagine the future of work.

To understand this confluence (between virtual and real worlds), one must first cogently define each of the two worlds, and subsequently look for facets that enable convergences, or divergences. The physical world has for long been clearly underpinned by three key facets – governments, economies and civil societies. Governments are responsible for establishing rules of behavior amongst the civil society so as to maintain and enhance citizen development while protecting their

Figure 1. Multi-nodal factors

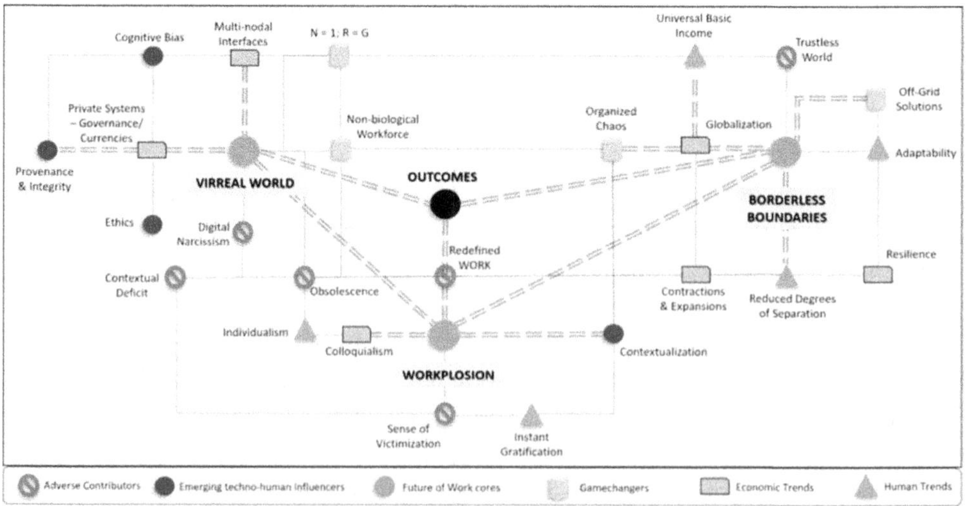

fundamental rights (to privacy, equality of opportunity, freedom of choice and speech, participation in economic endeavors et al). Civil societies – structured around common interests, origins and shared histories (culture, arts, language, and practices) – remain the source of effort, and target for consumption. The third facet – economics – is where the interplay between the other two is evident: markets are organized, production and consumption structures established, legal and regulatory frameworks determined, and cross-border exchanges undertaken.

World 1.0: Real World

For over two hundred and fifty years now, mankind has endeavored at building and sustaining societies based on certain orders, driven primarily by culture and civil preferences, and organized around constraining conditions (that may have their origins either in latent histories, past conflicts, and/ or tradition). Across the planet, such an organizational effort has given rise to a variety of models*. Some that have been consecrated to history, while others continue to thrive. For the most part, the world has adopted variations to the broad language of democracy, which in turn has given rise to the interplay amongst governments, private sector and civil society.

There are six key facets governing the physical world that in turn – combined – offer a veritable smorgasbord of opportunities and risks.

First, **governments** organized by the civil society to govern their behavior have continued to be structured around majority mutual consent (of the governed), translating into legal and regulatory frameworks that in turn act as guidance for orderly establishment of activities and exchanges, both economic and social.

Second, **economies** have been created through structuring productive endeavors that in turn provided gainful opportunities for participating societal members. Exchange of effort for value continue to underpin economies worldwide, where societal needs are met. A range of participants have joined forces to deliver requisite value as perceived by the citizenry.

Third, **culture and religion** have continued to drive a range of acceptable and unacceptable behaviors from all participants – citizenry, corporations and governments. Interactions amongst people within an economy/ nation state are broadly homogeneous[†]. While some distinctions are made within the context of culture, in most cases economic models tend to remain largely impervious to them.

Fourth, **production ecosystems** have been established on the premise of the greatest good for the greatest number, meaning that what is good for one group must be good for others too. Mass customization of products, large-scale production of goods and services, homogenized approaches toward building and growing production systems has always been premised on continuous consumption, supported by regulated structures toward competition, value assessments and fungibility of money. Consumption ecosystems have marched hand-in-glove with production ecosystems, thereby creating positive-sum game opportunities for all.

Fifth, **fiscal and monetary systems** that traditionally premised themselves on the maxim "your total value is the sum total of your owned assets[‡]" were upended after the second world war, leading to the creation of a new system known as fractional reserve banking. Separation of fiscal and monetary responsibilities reflected the maturity of a nation-state's governance structure, notwithstanding the fact that this system inherently created opportunities for humans to overplay a given hand[§].

Sixth, **world trade** that was earlier defined by rampant colonization, asset extraction and slavery, translated into a more orderly behavior among nation states after the excesses of the two world wars. Formation of the United Nations, and subsequently a range of global finance and trade organizations like the IMF, World Bank, ADB, WTO (format GATT), transnational governance entities like the ICJ, ITU, GSMA, other trade groupings like EU, AU, ASEAN, CARICOM, and finally inter-governmental free-trade agreements (both bilateral and multilateral) have collectively contributed to the nature and growth of global cross-border collaboration**.

The world has come a long way from its ancient history governed by kingdoms, serfs and servitude. Meanwhile, over the past century mankind has endeavored at significant value creation, through continuous discoveries across a range of disciplines. It may be a safe to state that our species has come a long way, more so in the past five decades that it ever did in the past five centuries. Interestingly however, this world order we have created has now – seemingly – reached a point of maximum utility[††]. Questions are being asked about the continual efficacy of our systems of governance, trade, utility and globalization as a plethora of issues surrounding marginalization, exclusion, poverty, modern slavery, economic discontent, and corruption continue to plague us.

World 2.0: Virtual World

For better or worse, humankind seems to have always endeavored at enhancing its lot each new day. Ingenuity and boundless curiosity has continued to underpin our actions as a species, so much so that we seem to be reinventing ourselves continuously. A telling example of this behavior manifests in our maturity where the entire world order is being questioned for its efficacy, and utility. Examining these from a standpoint of the six underlying forces that govern the real world is pertinent (Varanasi, 2018).

First, **governments** continue to be structured around collective representation of a nation state's civil society. Yet, the underlying structures, consent-based decision-making, and checks and balances defined by law and judicial prudence, are being upended. Citizens are beginning to chafe at the restraining behavior – overtly and subtly – being imposed on them‡‡. The consequences of initiating such strictures may not be understood by many. However there is continual discontent among the intellectual elite surrounding extant intransigence. In a highly networked and virtually connected world, it may not be possible for individual freedoms to survive the onslaught of forced homogeneity as much as the lawmakers may want to impose.

The rise of far-right wing ideologies across the world – both developed and developing – is a telling articulation of the failure among governments to act as mere representatives of the governed. In the absence of choices, citizens continue to push back – with some successes – on limiting environments, thereby making for a veritably complicated and "in-the-moment" implementation of governance. The absence of a long-term focus[1] has further added to the perceived misery of citizens in the context of declining lifestyles, ambiguity with their interim futures, job losses and significant sustenance woes.

As inequality increases alongside wealth gains, discontent is growing everywhere. Considerations around building universal basic income (UBI) systems[2] are beginning to gain traction, so as to alleviate the woes of people on the margins of poverty and development.

Second, **economies** hitherto built on the pedestal of consumptive behavior and resource extraction that supported human needs, are being questioned for their inability to create inclusion. Poverty is widespread, inequality is on the rise, yet the controlling levers to shape economic direction seem to be confined to a minority of successful businessmen. The control such individuals hold over an economy are being increasingly questioned by nimble and agile youngsters, more so in the form of creating startups that offer alternative solutions to traditional problems. Exchange of effort for value – considered sacrosanct for a long time – seems to have become secondary to the immediate personal gains of business owners. The dwindling distribution of the "gains" made from such efforts have created significant angst amongst the working populace. Further eroding the trust is the upending of business and operating models around the world.

Technological innovations have been able to not only eliminate inefficiencies from a hop-step-jump system of delivery, but also disintermediarise long-standing businesses that worked within supply chains connecting primary producers to end consumers. Increasing unemployment, job losses, and significant underemployment[3] may not only have reduced discretionary consumption, but also changed the view such individuals held toward their existing economic systems.

Compulsory spend – on basic needs – seems to be on the rise despite technological innovations. This has given rise to a small (but increasingly compelling argument) for "off-grid" solutions and adoption. For e.g. solar panel installations and connected homes are helping small communities / micro-societies take themselves off the electricity grid, and instead sell excess production (of power) back to the grid. This conversion of consumers to producers is quite telling. Similar stories abound as it relates to micro-farming and self-sufficient satellite communities around the world. The consequence of such adoption is that society is increasingly disassociating itself from its government, its elected representatives, and by extension, the "common" rule of law that keeps chaos at bay. Are we heading in a direction where economic theory and practices need to be fundamentally revisited?[4] Will nation states will cease to exist the way we know them?

Third, **culture** has for long been driven by mutuality of behavioral code within a community, and shall continue to remain so. Morphing traditional value-systems to suit contemporary norms is not easy (as evidenced globally with various movements like acceptance of LGBT community, gender orientations and rejectionist behavior of various communities). Technology may only be able to influence culture partially. The bigger influence manifests in human mobility. Countries today ready themselves for the onslaught of heterogeneous behavioral patterns, while simultaneously protecting their core heritage[5]. This readiness translates into new (and mostly untested) conditions – legally and otherwise – for all participants. Time is the most crucial factor here that can permit reflection on whether a certain readiness element has been progressive or regressive.

Alongside, a sense of continued ambiguity, lack of predictability and constant "orderly chaos" has begun to push people toward absolutes where constancy is guaranteed. **Religion** stands steadfast in providing a needed assurance that while things are rapidly evolving around, much still remains static. Fundamental overtones with religion have become commonplace globally across all religions today, and rejectionist attitudes abound, particularly around secularism, tolerance, and mutuality of respect. The virtual world has contributed significantly to distributing such belief systems worldwide, thereby enlisting a large and growing populace toward adopting rigid stances with policy or governance. Globalization is at its nadir today as it relates to religious and social tolerance. Meanwhile, the increasing number of non-believers, surrounded by a larger number of traditionalist believers are collectively leaving no room for a rational discourse.

Fourth, an area where technology has had the most profound impact is with the **production** ecosystems worldwide. Disruptions to traditional producer models, combined with complete annihilation of physical supply chains, distribution networks and scores of players that connected to consumers, has upended the view that jobs are a prerequisite to building a productive entity. Jobless growth in many parts of the world has become commonplace, while job losses too abound. Predictability is no longer possible for producer-consumers to imagine their immediate futures. In such contexts, organized / formal economies are slowly languishing, while informal economies – the bane of economic losses – are regaining traction. Implications are clearly manifested in a dwindling direct tax revenue base. Countries struggling to fund activities are continually increasing indirect taxes, thereby overburdening all citizens. The snowballing impact is an increase in income disparity, which has significant political implications for nation states. Rejectionism, anti-immigration, anti-trade (cross-border), communal and other forms of separatist behavior are manifesting non-linearly.

While technology has enabled greater levels of productivity and efficiencies, it has also become fairly exclusive for at least half of the world's population. "Digital divide" conversations seem quite meaningless, as they only focus on connecting people (yet not on enabling them to become contributing producers, and thereby active consumers). A larger piece to the solution is missing. This gap continues to fuel consternation, withdrawal and angst.

Fifth, the traditional **fiscal & monetary systems** in play worldwide, premised on "single sources of truth" have resulted in an unfettered pursuit of wealth. Predatory behavior with lending and credit systems have for long fueled exclusionist behavior among both producers and consumers, leading to significant wealth accruals unsupported by broader distribution. Controlling entities like central banks and regulators have often erred on the side of producers rather than the consumers. This imbalance has reached enormous proportions, tilting global financial systems on its head. Ingeniously crafted products and services have collectively led to fiscal failures alongside an acceptance that deficit financing models are the only way for economies to prosper. Money creation and supply, resting in the hands of the few and controlled by large producer cartels has further eroded our collective trust in the fractional reserve system of banking.

Concurrently, technological innovations in the Fintech world are rapidly – albeit discretely – breaking down the stranglehold of traditional players on economic and financial systems worldwide. A smorgasbord of solutions today vie for space alongside banks and financial institutions. They are slowly chipping away at long-standing norms toward lending and credit, exchange and value assessments, transactions and trade. While many such new players are pushing the gauntlet and seeking out customers, traditional institutions are scrambling to retain their consumers (and consequently indulging such new solutions, with both awe and consternation).

The global monetary system itself is coming under serious pressure. Currency trade, and the traditionally held belief that fiat currencies are the only way to properly establish value in exchange, are being discarded. Virtual and crypto-currencies seem to offer a comprehensive (yet untested) solution to our collective woes. The premise that complete elimination of "single sources of truth" was the only way to build a new future where equity could be assured, has taken firm roots.

Other solutions like private currencies, tokens, stored value devices/ cards are beginning to replicate ancient exchange (barter) systems, where participatory behavior and ubiquity with acceptance / rejection of transactions or attendant value are seen as the most proximate yet poignant alternatives to fiat currencies. Young consumers have easily adapted to this innovation, while traditionalists – governments, banks, institutions, regulators – continue to be wary [6]. The bigger battle here would not be around which system wins. Instead, we believe it may be around the modalities surrounding co-existence.

Sixth, **world trade** hitherto premised on access to markets, products and services in exchange for the same has begun to take a big beating. Continuous revisions to trade agreements has become the norm (rather than the exception) given that the virtual world has more or less eliminated national borders to production and distribution. Alongside, global agreements involving patents, intellectual property rights and related aspects are continually being challenged, to the point that erstwhile protections do not apply any longer.

Notwithstanding plagiarism, there is little room nowadays for organizations to sustain a monopoly or supremacy over a product or value proposition. Technological innovations have again permitted seamless adoption, adaptation and localization of ideas and solutions, much to the chagrin of policy makers. Domestic producers in most nation states have begun to revolt against the tirade of unfair competition from around the world.

Responding to such woes, governments have begun to either renegotiate or completely walk away from traditional trade deals[7]. The consequential impacts to markets, and consumptive behavior are yet to be properly understood. Traditional cartels like OPEC languish on the sidelines, while entities like the G7, APEC, OECD and the UN are being replaced by entities like the SCO, G20 etc., all premised on greater participation from the developing world[8].

Conundrum in a Virreal World

The world as we know it is being increasingly questioned for its overall efficacy and utility. While we have endeavored for decades on making improvements to the manner in which our socio-political and econo-commercial ecosystems collaborated and created value, the very premise underpinning these ecosystems was never in question, since its acceptance more than seven decades ago.

However, we have reached that point in time, where enhancing the old with the new is no longer seen as sufficient. The belief that the virtual world is a more efficient and modern extension of the real world is being debunked. Instead, the virtual world has almost completely discarded the practices of the real world, and is building the foundations for the rise of a potentially new world order.

The implications to such an order are profound. From a standpoint of how the future of work would look like, none of us have a handle on either the direction, or the goals. We are currently in the blind, more or less swinging with the tide as changes are unfolding. A very small minority of individuals worldwide have a sense of tomorrow, while the majority, including organizations and governments are in the dark about actions to take that could help sustain that tomorrow we are heading toward. Defining, redefining, reinventing or reinvigorating our approaches toward growth, sustenance, effort and value creation have become increasingly ambiguous, and transient. This pertinent situation shall continue for quite some time before any formal structures are established. Meanwhile, it is crucial to remember that the new does not replace the old; instead, it only layers itself on top. Accepting this reality is perhaps the only way we can make sense of the future of work, and all other attendant aspects governing civil societies and nation states.

WORKPLOSION AND BORDERLESS BOUNDARIES

Structural changes in the workforce can be painful, but ultimately progressive as we have been witness to over the past two centuries. The nature of such changes have – in almost all instances commencing with the first industrial revolution – taken the world by storm. Dissent and acceptance followed, inevitably leading to modified and enhanced adoption over time. Value derived from such changes may have, for the most part, been considered incremental. However the impact of such changes have been nothing less than profound for most of humanity – both positively and adversely. We are in a continuous cycle of such change that one doesn't look into it in too much detail. Most initiatives are considered progressive in the long run, hence the need for bemoaning short-term adverse situations or conflicts are not viewed with much care and attention.

This cycle has yielded positive results, as can be evidenced from the nature of human capital management in organizations world over, where structure and discipline, intermixed with cultural and local nuances have been successfully embedded into management practices, while maintaining growth and progress. The latter have become lagging indicators to the success (or failure) of any management practice worldwide. Books have been written, speeches made, theories propounded, mindsets changed, and corporate clones created. Any and all significant progress made by organizations has been variously touted as the result of collective endeavor, or team-work and camaraderie, progressive policies or meritocratic management style. We have imbibed this logic without much question, given that the benefits outweigh the collateral damage that may have been caused[9]. One cannot deny the positive numbers on a balance sheet now, can we?

Factors to Production

A range of factors influence production, particularly in times when a globalized world confluences – positively or adversely – with integrated supply chains, access to remote labor, cloud-enabled technologies and competencies. Dependence on fixed-cost input factors is being quickly replaced by "circular" factors, governed by re-usability and fungibility of such factors. The advent of an era where industrial dependencies are being upended has permitted organizations to relook at their business models from two contexts: (a) ongoing sustenance, and (b) future relevance. Today's capitalistic approaches continue to influence management behavior toward capital and resource allocations, much to the chagrin of contributors that could as easily replace / enhance value to products and services, if only business models could accommodate. In addition, one is witness to the continual demise of business models languishing in the past as their internal rigidities go against the grain of constantly shifting consumer patterns. Obsolescence and irrelevance are fast becoming a norm, rather than the exception.

The context with traditional factors of production as being less relevant today than before has yet to sink in. Most economies are not ready for such fundamental shifts. A slow, steady and linear progress toward modern agile industrial sectors does hinder the ability to build domestic resilience. Dilution in readiness and wherewithal to absorb external shocks is apparent across much of the developing world. The eventual consequence is a push-back on globalization in its general context, manifested variously in policies, tariffs, cultural rigidities and attendant lackluster competencies. **While information has become extremely fungible, borderless, and accessible, knowledge and applicability continue to remain a slave to economies that have preserved resilience.**

These non-linear factors influence production across a range of economies adversely, and spillover into a distinguishable gap between provisioning and expected returns. One principal impact is felt with "human capital" where skills and knowledge are fast being eroded in favor of technological excellence, and replaceability. While most economies have initiated strategies around resource reskilling, much of their effort is limited in its value, owing to presence of a significant amount of uncertainty with predicting future trends. This situation does present more choices than one is willing to take, for the nature of value being pursued is largely unknown from a standpoint of sustenance or continuity. Jobs therefore continue to hold-on to traditional guarantors, confined to command-and-control environments, while value erodes. The future of jobs is quite at stake within organizations going forward.

Tryst With Knowledge Obsolescence

It's been over a century since mankind has explored, discovered and invented across a wide spectrum of activities leading to a direct impact on human progress. Notwithstanding conflicts between science and theology, or between ethics and cultures, mankind has sustained its knowledge and grown it continually. Much of humankind today leverages such knowledge to produce and consume. The interesting element with such knowledge is that its creation has dramatically changed over the past two decades than in any other time in history. People from previous generations leveraged knowledge gained from education systems for a significantly long period of time (approx. 35 years for those attaining adulthood after WW II), without the need for revisiting their competencies. Fast-forward to today's day and age, knowledge acquired from educational pursuits are becoming irrelevant within a much shorter span of no more than 2 years. This phenomenon leads to continual upheavals with an individual's perceived capabilities in the context of applicability. The need to continuously unlearn and relearn is putting significant pressure on individuals, leading to many exclusions within communities.

One may argue that technologies are enabling, however that's missing the forest for the trees. An individual's worth in a community, in an organization and in society have been, and continue to be assessed by the individual's overall worth – or contribution to economic endeavors. The exclusion is throwing a large spanner in the works, leading to significant woes within people. Most manifested are shifting behaviors toward colloquialism, rising resentment, victimhood, and the like. Meanwhile, what is observed in general revolves in contextual deficit, low attention spans, digital narcissism, and moral conflicts.

These factors have together contributed to significant woes for economies, to the extent that policies are being pushed in the direction of exclusion, while soft barriers to open trade are being increasingly propped up. Agreements like WTO, WIPO et al are being increasingly questioned for their worth and spirit of inclusion. Liberalism is fast being replaced by right-wing thought, fueled by despair and an inability to deal with the ambiguity and uncertainty of immediate short-term futures.

Human endeavor at leveraging technologies significantly are contributing to an uneven distribution of competencies. Modern day slavery is a thought-process governing most young individuals who are choosing to disassociate from established governance systems, calling for fundamental overhauls. What shall give, and what shall remain is anyone's guess. Nevertheless, the interim victim culture permeating economies and civil societies must be addressed proactively by both governments and the private sector.

CONCLUSION

Its indeed a complex world. The three briquettes addressed in the pages above are but placeholders for the varied intertwined nodes influencing human conditions. As we continue to pursue progress, and drive toward greater inclusion of people, societies, industries and economies, we are also contributors to the attendant discord caused by such pursuits. Acknowledging their impacts and outcomes is the first positive step toward envisioning a better, and saner future where humankind co-exists with nature; science compliments endeavors; reason governs actions; collaboration triumphs over individuals; purpose guides solutions; and sustenance leverages collective accountabilities. The change shall continue to remain constant, while agility and flexibility with empathy shall help us build a future that does not enslave many for the benefit of some. Altruistic though it may seem, there is enough proof that a collaborative effort does indeed enable our societies.

REFERENCES

ADB – Asia Development Bank. (n.d.). Retrieved from: http://www.adb.org

ASEAN – Association of South East Asian Nations. (n.d.). Retrieved from: http://www.asean.org

AU – African Union. (n.d.). Retrieved from: http://www.au.int

CARICOM – Caribbean Community for Regional Integration. (n.d.). Retrieved from: http://www.caricom.org

EU – European Union. (n.d.). Retrieved from: http://www.europa.eu

GSMA – Global System for Mobile Communications Association. (n.d.). Retrieved from: http://www.gsma.com

ICJ – International Court of Justice. (n.d.). Retrieved from: http://www.icj-cij.org

IMF – International Monetary Fund. (n.d.). Retrieved from: http://www.imf.org

ITU – International Telecommunications Union. (n.d.). Retrieved from: http://www.itu.int

Muellj, D. (2010). *Redeeming Economics*. ISI Books.

Varanasi, R. P. (2018). *Enterprise Agility in Today's Era of Complexity. In Improving Business Performance Through Effective Managerial Training Initiatives*. IGI Global International Publisher.

Varanasi, R. P. (2019). *Future of Jobs and Humans in a Tech-Centric World. In Forecasting and Managing Risk in the Health and Safety Sectors*. IGI Global International Publisher.

World Bank. (n.d.). Retrieved from: http://www.worldbank.org

WTO – World Trade Organization. (n.d.). Retrieved from: http://www.wto.org

ENDNOTES

1 A range of "organizing models" have emanated – democracy, socialism, fascism, Nazism, communism, theocracy, dictatorships. Some continue to hold sway over large swathes of humanity, while many have been rejected as unsustainable ideologies. This paper will preclude that readers have sufficient understanding of these models.

2 We are refraining from delving into specific details that drive micro-behavior amongst societies. A range of factors influence such behavior – race, creed, origins, migration trends, religious affiliations, belief systems, et al as these facets do not lead to a significant deviation from the broad principles of a nation-state's cultural and social stance.

3 The Gold Standard that governed world's total value was abandoned at Bretton Woods, New Hampshire in July 1944. The Bretton Woods Conference, officially known as the United Nations Monetary and Financial Conference, was a gathering of delegates from 44 nations that met from July 1 to 22, 1944 to agree upon a series of new rules for the post-WWII international monetary system. This system came to be known as the Fractional Reserve system that is currently in play globally.

4 The most consequential example of this issue has been the latest economic crisis of 2008, where enhanced risky models involving financial engineering and securitized lending translated into highly predatory behavior, culminating in a large-scale wipeout of intrinsic value. For the first time in history, humans have begun to question the very foundational principles of our financial and banking systems.

5 IMF (2021); ADB (2021); WTO (2021); ICJ (2021); ITU (2021); GSMA (2021); EU (2021); AU (2021); ASEAN (2021).; CARICOM (2021).

6 Maximal utility is defined by economics that, when making a purchase decision, a consumer attempts to get the greatest value possible from expenditure of least amount of money. His or her objective is to maximize the total value derived from the available money. Extending this logic to social and governance

conditions existent today, we may have reached maximum utility with our entire way of life and the world order we live in.

7 A case in point is the ubiquitous monitoring of citizens through facial recognition technologies by the Chinese government, and consequent implications on the credit scores, and personal freedoms of citizens. Other nations – like the UK - have implemented some form of citizen monitoring systems, mostly limited to security amidst terrorism concerns.

8 Policy making for the long-term has taken a big hit globally, resulting in shortsighted approaches toward development. Consequently, people languish at the extremities of economic endeavors, or are seeing their value/ contribution decline to the point that their collective discontent has the ability to organize rebellions, and at worse revolutions. The Arab Spring in the Middle East/ North Africa, crises in France surrounding liberal government, clampdowns in Hungary and Bulgaria in response to the rising tide of immigrants from Africa are just a few telling examples.

9 A variety of experiments with instituting UBI have come to the fore, either as small projects with controlled populations, or as potential game-changing mandates within nation states. Some earlier rejections in Finland and Switzerland do not explain the complete story; while new adoption and experiments in Kenya, Finland, California and India may yet offer a playing field.

10 World Bank's unemployment and underemployment indices are a telling indicator of this trend. Over 893 million individuals are considered to be in vulnerable employment (i.e. they are bound to lose their jobs soon). This does not include unemployed or unemployable individuals. (WORLD BANK, 2021)

11 *A compelling argument is presented by John D. Muelle (2010) r in his seminal book "Redeeming Economics". He argues that the original proponent of economics, Thomas Aquinas (1225-74) set forth four basic elements of economic theory – produce, consume, exchange, distribute. Five centuries later, Adam Smith sparked the second economic revolution by discarding two of these elements – consume, distribute. In 1870s, neoclassical economics finally realized the oversimplification by Smith, and restored one element – consumption (theory of utility). Perhaps we need to restore the fourth back again?*

12 *Telling examples of such behavior can be seen by nations (a) temporarily when hosting global events – corporate or sport – where huge overhauls are undertaken to satisfy the variegated needs of potential visitors; and (b) permanently, when permitting immigration and resettlement. Culture clashes are abound, yet managed prudently for the most part in a bid to adhere to the world order.*

[13] *The seemingly misunderstood behavior of crypto-currencies, fueled by the confusion as to whether they represent a "medium of exchange" or "store of value" have led many jurisdictions to clamp down completely, or highly regulate such currencies, and their trade. The anonymity of participants is completely anathema to the traditional "know your customer" model of self-governance instituted since the beginning. Either way, the rise of these currencies, tokens and coins is beginning to take on a serious undertone that they are not just passing fads, but here to stay. Some nations like Japan and Singapore are looking at ways to integrate such currencies into the traditional systems through fiat-pegging; while others are exploring options around running parallel systems. Meanwhile much of the world remains in a state of perplexity around governance, identity and security.*

[14] *Telling examples are (a) latest battle between the US and China surrounding tariffs; (b) Withdrawal of the US from the TPP, and the diluted rejuvenation as the CPTPP; (c) renegotiation of NAFTA; (d) increasing irrelevance of WTO; and many more.*

[15] OPEC – Oil & Petroleum Exporting Countries; G7 – Group of Seven nations; APEC – Asia Pacific Economic Cooperation; OECD – Organization of Economically Developed Nations; UN – United Nations; G20 – Group of Twenty nations; SCO – Shanghai Cooperation Organization.

[16] *At this point, it is pertinent to emphasize that this collateral damage has manifested by way of employees into "bell curves", significant disparities in wages tiered according to hierarchy, and also in terms of career progression strategies and executive compensations. The other significant – yet less talked about issue – surrounding wage parity between genders continues to languish on the sidelines. The author however would not wish to dwell into these "tactical" consequences as it relates to Workplosion. Instead, the emphasis shall rest on the larger picture influencing human behavior.*

Chapter 10
Computational Thinking and Didactics:
New Horizons

Maria Luisa Nardi
Independent Researcher, Italy

ABSTRACT

The failures and problems of school education today are caused by many factors, stemming from the bias of behaviorism that the teacher is burdened with, as well as being the result of changes in the way of thinking of new generations, often caused by the negative impact of technology – its source is usually in the process of teaching and growth itself. Computational thinking allows us to take complex problems, understand what the problem is, and develop solutions. We can present these solutions in a way that both computers and people can understand.

INTRODUCTION

One of the typical problems encountered in teaching, in particular in mathematics and other scientific subjects, is the student's excessive expectation of receiving specific instructions from the teacher to solve the problem - a step-by-step guide. Today's students, people belonging to Generation Z, expect immediate information, indicating how to solve the problem, along with quick results, are not prone to mental effort and have difficulty to relate to the reality of the tasks around them.

The expression "computational thinking" was first introduced in 1996 by Seymour Papert, while describing LOGO, the language of programming developed by MIT to teach programming for children. A decade later, in 2006 she was the scientist

DOI: 10.4018/978-1-7998-7126-2.ch010

computer science Jeannette Wing to take up the concept, defining it as a mental process that underlies the formulation of problems and their solutions, so that the solutions are represented in a form that can be implemented in an effective by an information processor, be it human or artificial, suggesting to look at thought computational as a fourth basic skill (together with reading, writing and calculating), essential for everyone, in everyday life.

Over the past three decades, numerous studies in Programming Psychology have been carried out. They are works halfway between computer science and cognitive psychology, with the aim of evaluating and improving the tools at their disposal based on the cognitive effects of these tools. Wanting to make computer science accessible to all, of particular interest are studies on misconceptions, which deal with finding concepts that are misunderstood by novice programmers, and studies on commonsense computing, which try to understand how people who have never received notions of programming and, more generally, of computer science, express (in natural language) concepts and computational processes.

In ancient Greece the concept of game assumed a specific characteristics. It was linked to the concepts of competition (Agón) and imitation (Mìmesis). Agón, or conflict, competition, competition, represented one of the ways of considering the game. The same Greek gods played with humans on earth to put them in physical or social competition with each other, and men created sporting versions of actual conflicts - throwing javelins, heavy stones, or arrows - to determine which individual or community had. the support of the gods. Mimesis, understood as the imaginary or dramatic staging by adults or children, included a disparate number of representational forms and was a precise form of worship aimed at imitating the gods. Imitative, interpretative or expressive, mimesis provided for acting out of the ordinary and was seen as an approach of human beings to the gods to seek their favor.

There are testimonies of the London of the thirteenth century that report the events of the elderly and young people, on the Sunday of Lent, gathering in the fields to fight for play.

During the Enlightenment and Romanticism, thinkers like John Locke, Immanuel Kant and Friedrich von Schiller resumed examining the human mind. The game was considered within the reflection on science, morality and art or as the foundation of rationality, or as the root of the irrational or spirituality.

The links between play and rational thinking led to the creation of detailed descriptions of what play should be for children and was elevated by educators and philosophers, such as Friedrich Fröbel, and even before that his teacher Pestalozzi339, to a activity carried out with playful and specific objects, suitable for shaping and modeling the body, mind, morals and spirit.

In the "children's gardens", that is the institutes founded by Fröbel in which the free growth of children and the acquisition of specific basic learning were guaranteed, play was considered a fundamental element of expression and knowledge. The same gardens were suitably equipped for playful activities and in them, in Montessori style, there were different types of games including geometric solids, which had the aim of promoting the intuition of the forms found in nature by recalling the essential structures of the world. In this sense, the game was aimed at the process of development and global education of children and considered as an "industrious doing".

Herbert Spencer, for example, revisited Schiller's notion of play, as an excess of energy, and transformed it into a psychological version of Darwin's ideas on adaptation. According to this reading, excess energy feeds instincts that help natural selection: fighting games were associated with the need for social domination; group games were associated with the need for social interaction; the artistic / aesthetic game was linked to the improvement of symbolic abilities.

Also the naturalist Karl Groos, among whose writings appear two classics on the game, Die Spiele der Tiere (The game of animals) of 1896 and Die Spiele der Menschen (The game of men) of 1899, interpreted the game in an adaptive function, to the light of Darwin's theory of evolution.

According to Groos, the so-called lower animals, those that have not evolved in the course of history, do not play; there is no evidence to support the belief that insects, fish, snakes, snakes or toads play.

It is also evident that in the higher species the young play more than the adult members; puppies play more than mature dogs, and human children play more than adults. This period of immaturity is characterized by the predominance of the game, and the longer the period of immaturity, the longer the play phase is. From these observations Groos derived that play has an adaptive purpose and contributes to the survival of the species: play during immaturity offers the opportunity to practice those things that would serve adaptive to the species during adulthood.

John Dewey in Democracy and Education recognizes that play is an imaginative, free and plastic enterprise. In play, unlike work, the interest is more direct, that is, the end of the game is its own activity, execution regardless of a further result.

The study of mental life has made evident the fundamental value of native tendencies to explore, to manipulate, tools and materials, to build, to give expression to joyful emotions, etc. When the exercises that are driven by these instincts are part of the regular school program, the student, as a whole, is busy; the artificial gap between life in school and life outside is reduced; reasons are given to pay attention to a great variety of materials and processes that are indeed very educational; cooperative associations providing information in a social environment are provided.

COMPUTATIONAL THINKING (CT)

The students' activities are emptied of basic and important elements of the knowledge construction process, such as: formulating hypotheses, deducing their consequences (including the acquisition of experience) and establishing any new hypotheses that take into account previous conclusions.

New technologies offer the ability to instantly confirm or deny a given hypothesis, so the student stops considering the correctness of the hypothesis, does not analyze possible errors, and uses another solution strategy without having made it his own. Passive and uncritical use of new technologies facilitates this process. It is in contrast with the method linked to the use of algorithms and programming, which instead require creativity, teach perseverance, and the ability to draw constructive conclusions from one's mistakes (Denning &Tedre, 2019).

Specifically, computational thinking, which accompanies the processes of solving problems with the use of computers, can be characterized by the following properties:

- The problem is formulated in a form that allows the use of IT methods and a computer or other devices for the automatic processing of information;
- The problem is in the logical organization of the data and in concluding them;
- The representation of the data is necessary both abstract, for instance in the form of a model or simulation;
- The solution of the problem is in the form of a sequence of steps, so it can be obtained as a result of an algorithmic approach;
- The design, analysis, and computer implementation of a problem lead to the most effective possible solution and the best use of computer resources and capabilities;
- The experience gained through solving a problem can be used to solve other problems related to it or in other areas with characteristics similar to the one solved. "

Computational thinking has been proposed by many as a fourth basic skill in addition to reading, writing and calculating. Given the relevant position in which it is placed, in recent years curricula, activities, frameworks, technical hardware and software tools have been proposed to teach computational thinking. The panorama, however, still remains very fragmented.

The problems with the term CT arise when it is considered not as an abbreviation to indicate "the scientific nucleus of computer science", but as the definition of a new discipline.

There is a deep difference between computer programming and computational thinking.

Figure 1. Computer programming

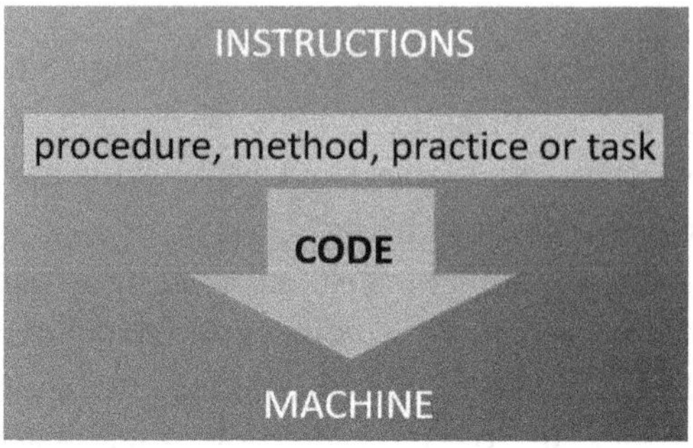

Figure 2. Semantic

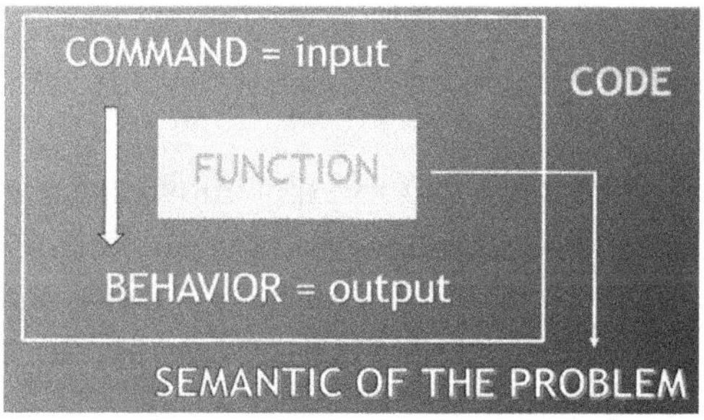

Programming: It means setting commands in a code. It requires memorizing a wide range of information and displaying multiple skills at the same time. A developer has to think about the details of the syntax to generate a statement.

Computational Thinking basically is a way of thinking that brings together various high-level skills and practices that are at the heart of computing, but applicable in many areas far beyond computing

Computational thinking is a problem-solving process that consists of: - formulating problems in a form that allows us to use a computer (in the broadest sense of the term, i.e. a machine, a human being, or a network of humans and machines) to solve them (Wang, 2016).

- Logically organize and analyze data;
- Represent data through abstractions, models and simulations;
- Automate problem solving through algorithmic thinking;
- Identify, analyze, implement and test possible solutions with an effective and efficient combination of steps and resources (with the aim of finding the best solution according to these criteria)

- Generalize the problem-solving process and transfer it to a broad spectrum of other problems.

A further difference is between Computational thinking and Coding.

Computational Thinking (CT) is the mental process that allows you to solve problems of various kinds following specific methods and tools; in a nutshell, the ability to solve a problem by planning a strategy. It is therefore a logical-creative process that allows you to break down a complex problem into different parts, more manageable if faced one at a time; by finding a solution to each of them you can solve the general problem.

This behavior is actually implemented every day, for example when we establish the shortest path to reach a destination or, more simply, when we play video games and we have to work out a plan to pass a level.

Coding is an English term which corresponds (approximately) in Italian to the word programming, or rather, its rudiments; in computer science means the drafting of a program, that is a sequence of instructions which, when executed by a computer, gives life to most of the applications we use every day.

The procedure for transitioning from the initial to the final state is described by an algorithm (procedure that solves a given problem through a finite number of elementary steps) that is performed by the processor (in our case, the individual).

What links these two concepts? It could be said that coding is the training ground for computational thinking that must be stimulated and trained from an early age. The most effective way to attract the attention of even the little ones is the play mode, that is the game: they, in front of a monitor, believing only to play, learn how to solve a more or less complex problem, writing a series of instructions the machine interprets and executes.

COMMON ELEMENTS OF COMPUTATIONAL THINKING

Computational thinking is a transversal skill. It serves to transform an intuition into a constructive procedure that leads us to the solution of a problem. Moreover, this solution will be so generic that it will be able to reuse that reasoning and that procedure every time we find ourselves facing a similar problem (generalization) Mental processes are computations on mental representations. In working memory there are mental models: they concern systems and objects with which we come into contact, serve to predict as such executables: we can do mental simulations on them.

Following the main sectors of CT development:

Mental Processes. They are mental strategies useful for solving problems, such as:

- Algorithmic thinking
- Logical thinking
- Breakdown of problems
- Abstraction
- Pattern recognition
- Generalization

Mental models are formed on the basis of the visible aspects of a system and their formation is inevitable [Ben-Ari, 2001], therefore it is good to explicitly teach the models that are considered correct. Visible appearance is the syntax. This however, especially at a high level, hides many details of the underlying performer model. When something is misunderstood, inadequate patterns and models are formed: we speak of misconceptions. Misconceptions in programming [Clancy, 2004] are due in particular to: syntax and notation (eg assignment seen as swap or increment as an impossible equation), English language (words like while, and, if with multiple meanings in English), attribution intentionality (computer with hidden intelligence that understands the programmer's intentions, computer with general vision, over-generalization of examples (I've never seen functions with non-numeric variables, I theorize that functions take only numbers as input).

Attitudes and habits:

- Research. Experimentation, free and open search for solutions, their comparison
- Creativity and imagination. Developing and using the imagination, proposing new solutions
- Improvements. A critical approach to the effects of one's work, focusing on the improvement and efficiency of methods
- Perseverance and patience. Withstand the pursuit of the goal, self-control in anticipation of the effects, awareness of the need to make an effort
- Cooperation. Work in groups and among peers
- Keep a real detachment from technology. Reflecting on the limits of technology while maintaining a critical attitude towards it

Methods. They are operational approaches used by computer scientists, such as:

- Automation
- Collection, analysis and representation of data
- Parallelization
- Simulation
- Evaluation

- Programming

Practices or Strategies. They are typically used in the implementation of IT solutions, such as:

- Experiment, iterate, do "tinkering"
- Test and correct errors (debug)
- Reuse and remix

Transversal Skills. They are ways of seeing and operating in the world; useful life skills fostered by "thinking like a computer scientist", such as:

- Create
- Communicate and collaborate
- Reflect, learn, do meta-cognition
- Tolerate ambiguity
- Persevere when dealing with difficult problems

SECTORS IN WHICH COMPUTATIONAL THINKING CAN BE DIRECTLY APPLIED

Computational thinking is appliable when it comes to interaction, processes, simulation, performance, behavior animation, modeling, prediction (Mcowan & Curzon, 2017).

Cognitive Science: Computational functionalism, as a trans-disciplinary study of mental abilities: language, thought, perception, emotions (and so on). Mental processes are information processing (somehow encoded in our brain), explaining the behavior of a system consists in reproducing it, in simulating it.

Biology: Genetics (genome sequence, genome assembly and gene prediction); Proteins (Modeling of protein interaction); Metabolism (Reconstruction of the metabolic pathway); Population (systems biology).

Chemistry: Discovery and design of drugs, Molecular dynamics simulations, Modeling of chemical pathways, Study of the fundamental behavioral properties of atoms and molecules.

Physics: Physical behavior of materials, Simulations of optical / sound performances, Astrophysical modeling, Physical interaction of biomolecules.

Medicine: Biometrics of productivity, Modeling of neural pathways, Simulations of physiological performances, Automation of medical surgery, Analysis of drugs and environmental toxins.

Engineering: development of mobile, internet computing (cloud), Gaming, Robotics, Computer and network security, machine and deep learning, and so on. This is because, for example, for the design of an expert system, a set of rules are formulated according to the methods of management, the mode of operation of the cause-effect relationship, of the principles that determine evolution. in computational thinking, the coding of the rules in function of a limited set of words with unique and well-defined meaning comes into play; and in the simulation of the set of rules according to the excursion of the force variables and the measured value of the output variables. And specifically: Agent Communities (Collaborative Problem Solving), Natural Language Processing (Nlp) System, Knowledge Engineering, Strategic Games.

Cognitive Tutoring: Simulation of user behavior (for a system: Profile driven User "Behavior Recording / Tracking"), Workflow Engine (static, flexible and dynamic rules), Instructional Design - Knowledge Management

Intelligence Analysis: Phases of an investigation process (preventive or abductive), Decisional phases, Phases of risk management, threats, criticalities.

Art: Image design, Movie animation, Face tracking, Mixed reality.

Music: Acoustics simulations, Recording techniques, Sound synthesis and manipulations, Computer music

Social Science: Demographics simulations, Pandemic reaction modeling, Computational finance

Education: Pedagogy of the Mind, Reality Tasks, Coding, Educational Robotics

COMPUTATIONAL THINKING AND TEACHING

Talking about CT within the school is a risky attitude: philosophers, rightly, ask themselves what we mean by speaking of "teaching thought"; mathematicians observe that many aspects of CT (algorithm, abstraction, recursion, problem solving, ...) are also typical elements of mathematics (which, however, is not called for this "mathematical thinking"); the pedagogues ask on what basis we are sure that the teaching of CT is really effective; teachers want to know what methods and tools are for teaching this new discipline and how they can learn to teach it; finally, parents are on the one hand happy because the school is finally aligned with an increasingly digital society, but on the other hand they are worried about what will happen to their children in the future if they learn to program with today's languages, which will be obsolete between a few years. Not to mention, then, those who would like to use the expression CT without any reference to the original article by Wing or to fundamental concepts of computer science, and who feel so free to attribute to it, from time to time, the most different meanings.

Some Didactic Tools

Blockly Games. An interesting application. It is a series of educational games that teach programming. It is designed for children who have not had prior experience with computer programming. By the end of these games, players are ready to use conventional text-based languages.

Twine. It is an open-source tool for telling interactive and nonlinear stories. Ultimately it is a digital storytelling tool. It is a program based mainly on written text and hypertext, but it allows you to create even more sophisticated products. This tool can find indications of use and applicability in contexts, let's say, humanistic or in general when you want to explore the interactive possibilities enabled by a digital system for the construction of narratives and reading experiences alternative to the usual, traditional, sequences of lines and pages. It allows you to completely invent an adventure game, a fantasy world in which you define characters, roles, challenges, objectives, antagonists, etc., and rules with which to conduct the game, acquire points, lose them, and so on. It stimulates creativity, imagination, literary analysis, teamwork skills, communication skills, expertise in the choice of adaptation of texts, images, audio and video, if you also want to have a multimedia experience, lastly but last but not least, the strict coding techniques, with which you will have to model and implement the rules of conduct of the game and check its execution and content.

Minecraft. The game is based on placing or destroying "blocks", that is, the cubes that make up the worlds in which the game is set, and on making objects by interacting with them.. The graphics are deliberately pixelated and similar to Lego blocks, allowing you to create worlds by choosing tools and materials and to meet more or less curious and disturbing creatures.

Scratch. It is one of the most popular tools is Scratch (to date version 3.0), a software dedicated to programming, a free environment that uses a graphical, block programming language, designed at MIT in Boston. There is even a "junior" version for those who still can't read (from 5 years old). With Scratch it is possible to program interactive stories, games and animations by sharing the products made with the members of the Scratch community. Very suitable for the development of computational thinking Scratch teaches children to think also creatively, to reason systematically and to work collaboratively.

Logo. It is designed to help kids learn programming hands on. Instead of memorizing theory or using complicated programming structures, LOGO users learn programming basics with simple words and directions. It works from a "soft" use of information technology to a "hard" use. Robots, electronics (Arduino), 3D printers are back in fashion.

Figure 3. Blockly games (https://blockly.games/)

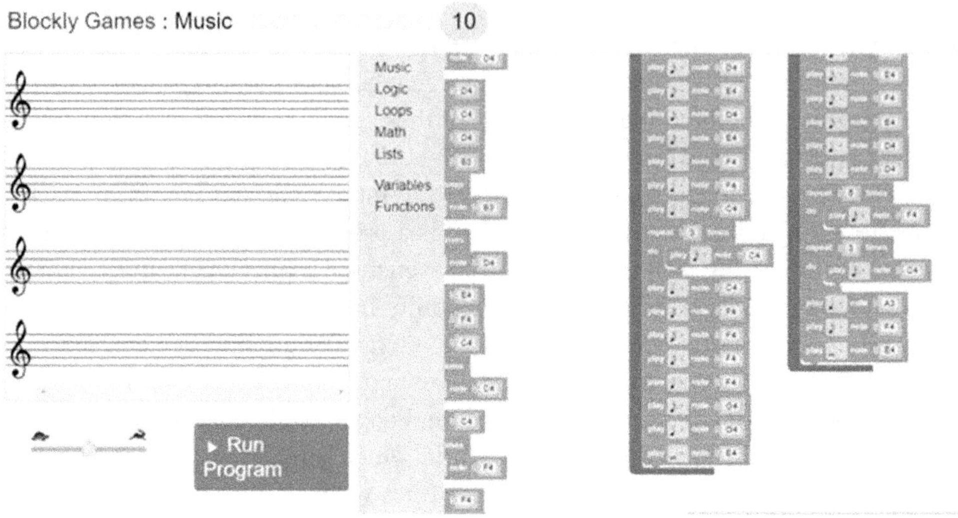

Figure 4. Twine (https://twinery.org/)

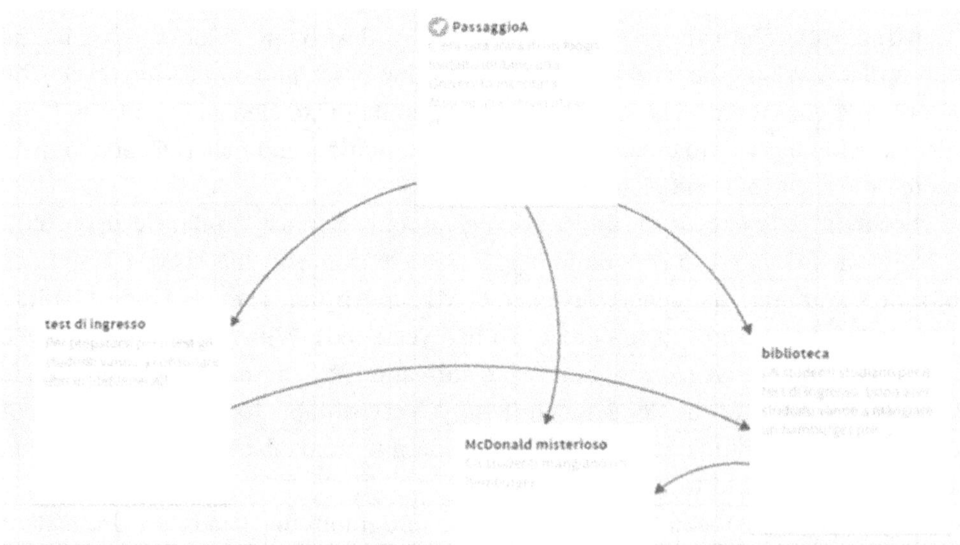

Figure 5. Minecraft (https://www.minecraft.net/it-it)

Figure 6. Scratch (https://scratch.mit.edu/)

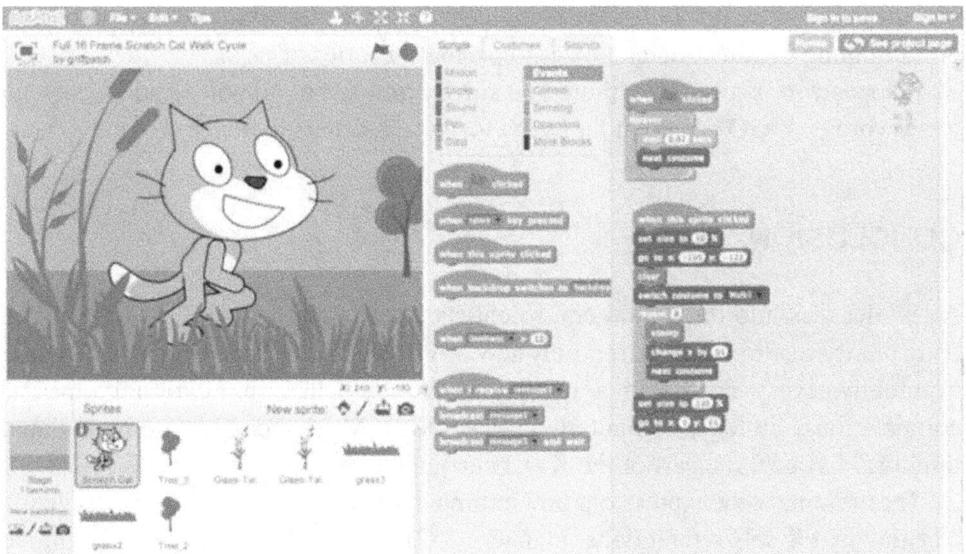

Figure 7. Logo Programming Language (https://people.eecs.berkeley.edu/~bh/logo.html)

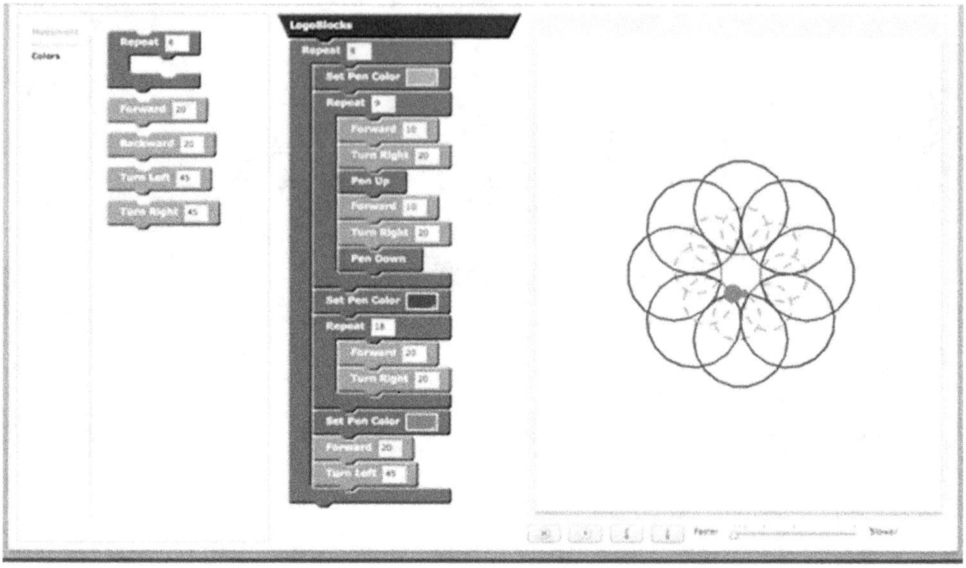

While LOGO began with the Lisp programming language and ideas about artificial intelligence, the goals of the language reflect the thinking of Jean Piaget, the Swiss psychologist. Piaget believed children learn best working in a constructive way, by building things. LOGO is a language that makes it easy for children to learn programming by directing an object to perform a series of tasks.

It appears to be used to explore and solve problems in graphics and physics. In other words, LOGO is a great language for kids, teachers, parents.

CONCLUSION

A few decades into the digital era, scientists discovered that thinking in terms of computation made possible an entirely new way of organizing scientific investigation; eventually, every field had a computational branch: computational physics, computational biology, computational sociology. More recently, "computational thinking" has become part of the K–12 curriculum.

The meaning of the expression "how a computer scientist thinks" can be sufficiently clarified by the self-referential explanation "CT is the set of mental and cognitive skills obtained through the study and practice of computer science": the defined "tacit knowledge" by Polanyi (Héder & Paksi, 2018).

In synthesis, computational thinking (CT) is not a set of concepts for programming; it is a way of thinking that is honed through practice: the mental skills for designing computations to do jobs for us, and for explaining and interpreting the world as a complex of information processes.

REFERENCES

Denning, P., & Tedre, M. (2019). *Computational Thinking*. The MIT Press Essential Knowledge Series.

Héder, M., & Paksi, D. (2018). Non-Human Knowledge According to Michael Polanyi. *Tradition and Discovery*, *44*(1), 50–66.

Mcowan, P. W., & Curzon, P. (2017). *Power Of Computational Thinking, The: Games, Magic And Puzzles To Help You Become A Computational Thinker*. World Scientific Publishing Europe Ltd.

Wang, P. (2016). *From Computing to Computational Thinking*. Chapman and Hall/CRC.

Chapter 11
Models for Strategic Thinking in Intelligence Analysis

Luisa Dall'Acqua
University of Bologna, Italy & LS TCO, Italy

ABSTRACT

Intelligence is a field of activity that has a rich history, yet it still lacks clarity of definition and agreement. This is not to denigrate the performance of many of its practitioners; rather, these comments are designed to point up the fact that, despite the length and breadth of its historical practice, there is still much to do to explore the boundaries, opportunities, and limitations of the application of intelligence in the world of enforcement. The cultural roots of strategic intelligence is compelling and urgent not just for those involved with the professional intelligence community, but also for anyone seriously committed to interdisciplinary studies, cross-cultural understanding, and most importantly, to the development of a rigorous discipline of politics as well as cultural genetics. But strategic planning does not guarantee strategic thinking.

INTRODUCTION

In the nineteenth and twentieth centuries, intelligence practice evolved toward the form that is now recognizable in the context of current practice. To its detriment, the intelligence milieu is somewhat misunderstood by many observers.

The concepts of espionage and critical analysis are often confused as if they were all part of the same activity. *Espionage* is about gathering data in the intrusive and invasive environment of spying.

DOI: 10.4018/978-1-7998-7126-2.ch011

Intelligence and analysis is a wider process of problem solving that involves data gathering and analysis, interpretation, and speculative consideration of future developments, patterns, threats, risks, and opportunities.

Strategic intelligence analysis can be considered a specific form of research that addresses any issue at the level of breadth and detail necessary to describe threats, risks, and opportunities in a way that helps determine programs and policies.

The role of Intelligence in strategy formulation is a critical one, especially now, given the quickly changing business environment. As the pace accelerates, it becomes more difficult for senior managers and decision-makers to adequately monitor, interpret, and respond to environmental changes. This is where SI, a systematic process of collecting, analyzing and communicating actionable strategy-oriented business information, can serve as input in strategic decision-making.

From the process perspective, Strategic Intelligence can be seen as a part of Competitive Intelligence, as it is also a systematic and continuous process with a purpose to facilitate decision-making with needed and timely intelligence input delivered in actionable form. The major difference between SI and CI is that SI actions focus strictly on supporting strategic decision-making by monitoring aspects with strategic significance. Additionally, a SI time horizon is broad and the focus is on all significant events: past, present and future events.

The code of practice for strategic intelligence has been fully developed and successfully tested for several years now. Like most routines and protocols, it needs constant maintenance and fine-tuning, but the reality is that the process works well. It sets out a highly disciplined approach to crime research.

Whether it is on particular groups, laws, or criminal behavior phenomena. The approach defines every element of the activity necessary to undertake a form of "conceptual" analysis. It demands that practitioners accept the need for highly disciplined and orderly processing of data, whether hard or soft, yet at the same time encouraging the development of intuitive and creative thinking to enable a high degree of reasonable speculation about topics that are often vague and data-poor.

What are the belligerents' survival, vital, major, peripheral and other diplomatic, economic and military interests in the region? What are their specific interests in this problem? What is the specific U.S. interest involved in this problem? How does this problem affect American survival, vital, major, peripheral and other diplomatic, economic, and military interests? What is the specific threat to U.S. interests? Is this a conflict between the United States interests and the interests of some other, significant, third party? How will this conflict affect each of the interests of the belligerents? To what extent do the interests of others compliment or conflict with U.S. interests?

Success Criteria. What is the definition of success? Is there a positive correlation between operational objectives, enemy strategic centers of gravity and the strategic objective? Does the selection and attainment of the operational objectives attack and destabilize the strategic centers of gravity or otherwise achieve the strategic objective? Actions Necessary to Achieve the Objective. What are the actions necessary to achieve the objective? Is the government organized to fight the war for the duration and throughout the spectrum necessary to achieve the objective? Has the government identified the political objective and prescribed actions necessary to achieve the objective? Exit Strategy. What is the exit strategy? What are the recognizable conditions for knowing when to exit? Is there a transition process identified? What are the military?

Demands on the National Will. What are the demands on the national will? Is there clarity of communication between the government and the people? Has the government clearly linked the underlying problems, affected interests and strategic objective in such a manner that it enjoys public support? Is there a strategy for adequately gaining and maintaining pubic support?

Demands on the National Capacity. What are the demands on the national capacity? Does the U.S. have the political leverage to obtain coalition support or is the U.S. confronted with a go-it-alone option? Is the political object worth going it alone? Are there sufficient resources for a go it alone option? Can the U.S. diplomatically afford to go-it-alone? Is the U.S. economic engine large enough to fight the war, underwrite allied support and sustain the public will? *Demands on the Military.* What are the demands on the military? Is the military doctrine, funding, resources and discipline sufficient to achieve the military objective? Is there sufficient political guidance and resourcing to develop and pursue a meaningful military objective? Can the objective be met with current doctrine and force structure? Are military resources sufficient in quality and quantity to achieve the objective?

These questions help Intelligence identify where the nation is strong, where it is vulnerable and what strengths may be available to compensate for identified weaknesses.

Political Leadership. Assess each belligerent's political trinity and the respective national will of each element within the trinity. What are the comparative strengths and weaknesses of each belligerents political trinity? Who are the key decision makers in each country and what has shaped their thinking? What is their likely perspective on this issue? How will this perspective be influenced by the role this individual plays in his government? How will the decision maker's perspective influence subsequent government actions? How does the governmental structure affect policy making and the decision makers and how can that structure be influenced? What societal influences exist and how will they affect the decision maker? What relationships exist between the decision makers' nation-state and the other international actors? How stable are the governments? Do relevant actors/belligerents have the power to influence the outcome and will they use that power to influence the outcome?

Are the military forces shaped and employed to initially control the four dimensions of the battlefield? How can the enemy most effectively interfere with that control? Is there sufficient military capability to achieve the objective including the desired end state? How can the enemy most effectively interfere with the achievement of that objective? What is the potential for mission creep as the mission develops? If there is a gap between the chosen policy and the means to carry it out, what is the best means for closing it?

It clarifies intelligence considerations for identifying and eliminating the gaps between policy and means.

We did write this however early on: "Life: A self-sustaining chemical reaction (or energy system) with built in expectations. Consciously or unconsciously it expects to be trusted if it does the one thing and deceived if it does the other. And to find trust if it does the one thing and deceit if it does the other. And to succeed by offering trust on the one hand or deceit on the other. All its calculating mechanisms are devised to predict consequences based on the probabilities associated with each of these expectations under all observable circumstances.

All evolution is the proximate result of the entity involved reacting strategically to its experience.

And in organisms adapting in turn to these mechanical tactics, suspicion of appearances (or distrust) also becomes the mechanism equivalent to suspicion of deception. Even though it will be ages before any of these organisms has the ability or need to see or deal with either deception or trust as an abstract concept.

So to leap forward a bit here, we might now define and more broadly describe all life as we know it as follows: Life: A self-sustaining chemical reaction (or energy system) with built in expectations.

Microorganisms continually redefine themselves at many levels, including the molecule, cell, and community. Although it was initially assumed that this resulted from the genesis of information within DNA alone, it has since been shown that innovation originates at multiple levels.

This occurs through calculative units, each unit consisting of two proliferating structures, one nested within the other and each undergoing changes in structural geometry that affect the proliferation rate of the other. For example, the recombination of genetic structures affects the proliferation of community structures, and the recombination of community structures affects the proliferation of genetic structures. The proliferation of a nested series of structures (e.g., genes proliferating within cells, cells proliferating within communities, communities proliferating within ecosystems) results in a logic circuit that calculates the form and function of each structural element in the series. In this situation each element functions as both a habitat and an inhabitant (environment and organism), and it is this dichotomy that determines the balance of nature. Nested geological structures, such as minerals and continents, also proliferate and

redefine themselves in much the same way. Microbe–mineral interactions thus link nested biological calculations to an analogous set of nested geological calculations. Examples include the microorganisms involved in the nucleation (proliferation) of ferric hydroxides, carbonates, silicates, and ice crystals.

WHY MANAGERS AND EXECUTIVES NEED STRATEGIC ANALYSIS

Strategic intelligence research develops in four ways:

- to inform managers, clients, and practitioners about strategic intelligence;
- to educate analysts about the processes and standards of strategic research;
- to encourage managers and executives to adopt the concepts outlined in the book because of what strategic research can do to help them achieve success, avoid failure, and minimize risk; and, finally,
- to show that the processes are simple enough of themselves, but need to be applied with robust intellectual discipline, creative flexibility, and openmindedness if they are to provide a successful outcome.

Basically, though, strategic analysis arguably calls for the same sorts of benchmark skills, qualities, and traits required of analysts everywhere. These core qualities include the ability to conceptualize problems, to consider and select the appropriate problem-solving tools, and to apply those techniques skillfully. Moreover, the analyst has to carry out this chain of activity with honesty and integrity, with care and attention to detail, being imaginative and creative, and all with discipline and orderliness in the way one behaves. These central qualities are basic to all analytical work in the intelligence field. Being able to apply them in those circumstances that demand different and particular skills and focus is what separates the various types of analytical functions (Mcdowell, 2008).

The pros and cons of conceptual model development are generally well understood, and, indeed, quite a lot has been written elsewhere on this topic. The nature of intelligence work, however, is such that it is often understandably accompanied by extreme pressures of deadlines to provide answers. While this

seems logical in the context of operationally urgent activities, running a strategic intelligence unit with the same pressures is arguably likely to become self-defeating, particularly if the time-critical targets are laid down arbitrarily rather than for reasons of common sense. In these circumstances, there can be strong pressure on the analysts to simply assume a level of prior knowledge about topics, avoiding what are seen as intrusive process steps that do not appear to have substantial and direct outcomes.

STRATEGIC INTELLIGENCE: ROLE, DEFINITIONS, AND CONCEPTS

Strategic Intelligence is a type of Competitive Intelligence produced for strategic purposes. Customers of SI include individuals involved in strategic decision-making.

The major difference between SI and CI is that SI actions focus strictly on supporting strategic decision-making by monitoring aspects with strategic significance.

Intelligence analysis must define the current competitive arena in which the company operates.

An intelligence assessment must be made of the forces and factors that make up the company's competitive environment.

The intelligence community throughout the world is still reeling after several intelligence failures. Proposals to improve Intelligence Analysis have had little impact as analysts, their managers and their organisations continue to cling to outdated threat perceptions, methodologies and organisational structures and cultures. This

Figure 1. The scope of strategic Intelligence
(Source: McDonagle&Vella, 1996)

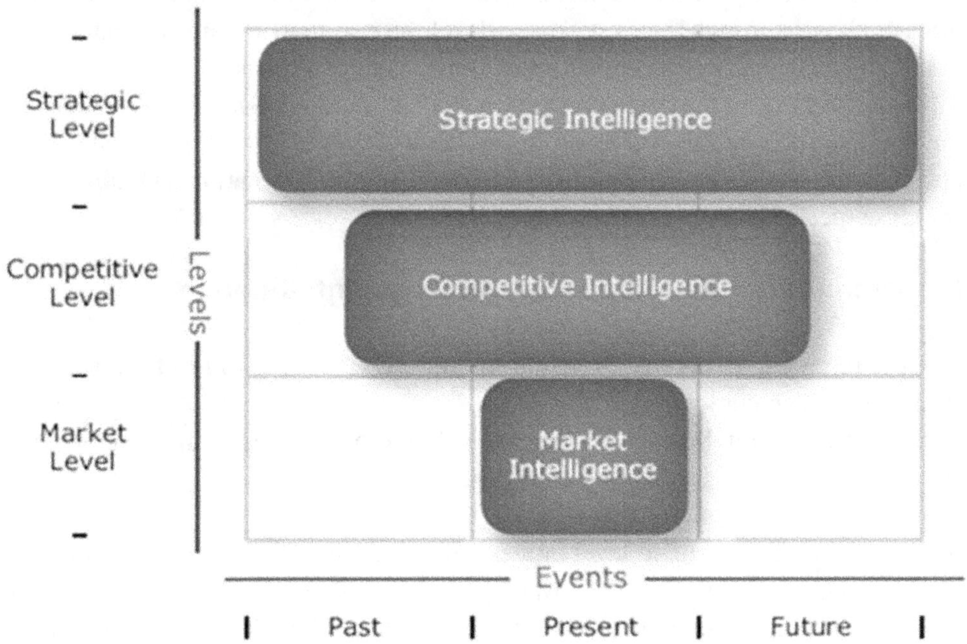

thesis looks through the lens of Knowledge Management at the various challenges that the Intelligence Analysis practice is faced with in the Knowledge Age. Firstly, theories and concepts from Intelligence Analysis are challenged when compared with those in Knowledge Management and the possibility of applying new vocabularies in intelligence is discussed. The second challenge intelligence

analysts face is to understand and adapt to the changed world with its connected, non-linear and rapidly enfolding events and patterns which broadens their scope to a multi-faceted, complex and multi-disciplinary threat picture. The third challenge is to re-look the existing analytical methodologies, tools and techniques, realising that these are most probably inadequate in a complex environment. The fourth challenge Intelligence Analysis faces is to reach out to other disciplines and assess how new analytical techniques, both intuitive and structured, as well as cognitive models, collaborative and organisational structure concepts from within the Knowledge Management discipline can improve Intelligence Analysis' grasp of the Knowledge Age. In conclusion, it is argued that intelligence analysts might be ready to reinvent themselves to address Knowledge Age issues, but that intelligence organisations are not able to support a new intelligence paradigm while still clinging to threat perceptions and structures befitting the Cold War (Duvenage, 2010).

The activities involved in strategic intelligence research are easily recognizable as part of the generic family of intelligence processes. Yet, because they have to be adapted to suit differences in focus and purpose, the overall approach and its process components have their own quirks and challenges.

Strategic intelligence and analysis practice focuses on being able to creatively think one's way through issues at a macro level, yet constantly retain pragmatic links to the inevitable tactical and operational impact and outcomes.

Even allowing for the depth of research that may be appropriate to some subjects, the activity always remains driven by a need to address just how the eventual results, conclusions, and recommendations can be anchored into operational readiness and response mechanisms.

Implementation and Impact of Strategic Intelligence

In order to practice effective Strategic Intelligence (SI) able to meet the needs of strategic management

and introduce intelligence of strategic value into the decision-making process, certain critical aspects should be considered, such as:

- Appointing a strategic level process owner, preferably the VP for Corporate Strategy, and a seasoned SI Manager enables integrating the best available internal senior expertise into the SI process. The SI Manager should be equipped with in-depth understanding of the battlefield and be respected as an expert in the organization's business and markets.

He/she should also participate in the strategy process.

- Knowing and understanding the internal customers. SI customers are individuals involved in strategic planning.
- Defining the SI priorities and properly communicating them (critical strategic themes, key players to be analyzed, early warning themes). It is important that the critical strategic themes be clearly communicated to everyone who is expected to contribute input to strategic analysis.
- Designing and producing SI products that serve the customers' content needs and communication needs. Here, "products" refer to highly interactive management briefings and visual actionable intelligence information packs as opposed to massive static reports.
- Building a network of Strategic Intelligence contributors, both internal and external.

Techniques for Strategic Intelligence

Following a taxonomy of strategies' techniques:

- Strategy as a *coherent, unifying and integrative pattern* for decision- making, a blueprint for the whole organization that indicates the organization's future destination.
- Strategy as a means of *establishing an organization's purpose* in terms of its longterm objectives, action plans and priorities for resource allocation; aligning short term action and resource deployment to be congruent with longer term objectives.
- Strategy as a *definition of a firm's competitive domain* including defining the businesses a firm is in or intends to be in, segmentation, strategic positioning and how and where it is going to compete.
- Strategy as a *response to external opportunities and threats* and to internal strengths and weaknesses; a means of achieving a long range competitive advantage, and a favorable competitive position and adapting to meet the demands of a changing environment.
- Strategy as a *logical system* for differentiating managerial tasks at the corporate, business and functional levels; defining an overall mission and allocating resources with a sense of strategic priorities.

- Strategy as a *definition of the economic and non-economic contribution* the firm intends to make to its stakeholders.

THE ROLE, RESPONSIBILITIES, AND FUNCTIONS OF THE ANALYST

Analysis has, in its own right, become a generic specialty within the field of intelligence. Insofar as intelligence activity is conducted in different environments that call for the application of highly specialized techniques, analysis itself is becoming somewhat functionally divided and compartmentalized. The language and culture surrounding analysis has undergone change, and, for example, job descriptions of analytical positions can vary significantly, as can the defined selection criteria.

The strategic analyst has several roles to play in carrying out an assignment. Moreover, analysts generally find that the "image" and status of intelligence and analysis within and beyond their organization is something that needs to be constantly worked on. The discussion that follows deals with the key responsibilities the analyst must bear as part and parcel of the notion of professionalism.

One of the hallmarks of strategic analysis, as with any other, is that the analyst is required to be able to think in leaps and bounds and not be circumscribed by precedent or conformity. Any reasonable observation of investigation practice would highlight the fact that the chain of logic and evidence must be impeccable if the result can withstand scrutiny, whether this is in the courts, the boardroom, or on the stock exchange. In contrast, while intelligence supports investigations in some settings, that does not make it a form of investigation practice. As mentioned elsewhere, intelligence relies on clever interpretation of data and events to identify what might be happening, what might occur next, who might be involved, and what could be the impact of such actions. This is a service provided to decision makers to allow them to focus on problem solving and opportunity taking. It is not an investigation with a finite, provable endpoint. History is the proof of good analysis (MSSI, 2002).

Risk Taking, Speculation, and Courage

Many agencies in Europe and North America have and continue to explore ways and means of making sure that the appropriate people are selected to work in strategic analysis. That some consistently assume that an academic education is the arbiter of performance potential is, in this current era of managerial thought, understandable. Nonetheless, one can easily find the grounds for argument that strategic analysis is, after all, a fairly simple set of processes that can be learned. It is the issue of personal traits and mindset that becomes more important in this discussion, and few agencies in my experience address this aspect comprehensively. In fact, most of the current models of intelligence standardization established back in the late 1990s still focus upon the skills-giving process and less, unfortunately, on the need to shape the behavior and thinking involved in their application (Niles, 2002).

The analyst's role in strategic intelligence research is just one of the many applications of analytical skill. In this particular setting, the analyst must be able to show a personal capability to undertake the type of research regime demanded.

Moreover, as has been stressed several times earlier, the strategic analyst has to accept the intellectual responsibility for "ordering" the entire research project to ensure that all the complementary elements—project design and data collection, for example—fall into place. While one might consider that this is a fanciful and unworkable concept, the past fifteen years of strategic intelligence practice in enforcement shows just how fragile and fraught the analytical work is when this leadership role is not allocated to the analyst and, for whatever reason, the notion of centralized intellectual control is sacrificed. To be wholly effective as the lead strategic researcher in a project, the strategic analyst must spare no effort to find the means of convincing managers and clients of the benefits that accompany placing "control" of the research project in its entirety with those who, from an analytical standpoint, run the project.

The responsibilities and obligations of a strategic analyst are not trivial.Apart from the workload volume stress that obviously goes hand in hand with this form of research, there is the additional burden of carrying the responsibility for conducting research on issues that may have such far-reaching impact

Strategic intelligence (SI) is an important tool in informing and supporting strategic management activities in different stages of the strategy development. In short, Strategic Intelligence can be defined as *"a systematic and continuous process of producing needed intelligence of strategic value in an actionable form to facilitate long-term decision-making"*. Customers of SI include individuals involved in strategic decision-making (dall'Acqua & Gironacci, 2020).

CONCLUSION

Strategic intelligence is a manager's tool. In the context of policing, for example, it is about decision making on important and overarching issues, not simply about targets for arrest. No manager can expect to function with full effectiveness if she doesn't have all the factual data; nor can one hope to properly proceed with enforcement unless there also exists a wide level of understanding of relevant problem issues. Strategic intelligence and analysis is key to providing this enhanced level of understanding, and it is always pitched at the level of command being exercised, an aspect of strategic product development not always understood by managers.

Senior executives in all walks of life, policymakers and lawmakers—including politicians—could gain benefit from the understanding that such insights, though occasionally unpalatable, provide real gain in illuminating the problems and providing both warnings of changed threat and forecast of potential opportunity. It is not uncommon for changes to come from the bottom up.

However, in the case of strategic intelligence, the experience of the past several years has tended to show that unless "management" is committed to acquiring strategic insights into problem solving, there is little likelihood of real change driven by analysts themselves doing strategic research.

REFERENCES

dall'Acqua, L., & Gironacci, I.M. (2020). *Political Decision-Making and Security Intelligence: Recent Techniques and Technological Developments*. Hershey, PA: IGI Global.

Duvenage, M. A. (2010). *Intelligence Analysis in the Knowledge Age- An Analysis of the Challenges facing the Practice of Intelligence Analysis*. Stellenbosch University.

McDonagle, J., & Vella, G. (1996). *New Archetype for Competitive Intelligence*. Praeger.

McDowell, D. (2008). *Strategic Intelligence*. Academic Press.

MSSI. (2002). *Strategic Intelligence Considerations When Preparing for War* (MSSI Thesis). National Intelligence University, Defense Intelligence Agency, Bolling Air Force Base, Washington, DC.

Niles, R. (2002). *The Strategic Intelligence of Trust Life, an Evolutionary Force of Nature*. CreateSpace Independent Publishing.

KEY TERMS AND DEFINITIONS

Intelligence: Is sometimes described as a "much abused" term in both scholarly literature and official discourse. This is in part due to the fact that national and institutional differences of perspective exist, complicating the search for definitions. Broadly speaking, intelligence can be defined in three contexts:

Intelligence as a Product: The third context in which intelligence can be defined is that of the product of these processes; a body of information and conclusions drawn from that which is acquired and furnished in response to the known or perceived requirements of a client. It is often derived from information that may be concealed or that is not intended to be available for use by the acquirer.

Chapter 12
Unmanned Aerial Vehicles:
Their Use During the American Presidencies of Bush Jr. and Obama

Giulia Casmiro Scarmagnani
Alma Mater Studiorum, University of Bologna, Italy

ABSTRACT

Last January, the General Secretariat of the Council of the European Union produced an estimate that, by the year 2035, the drones' sector will create up to 100,000 new jobs and, in addition, will have an economic impact of more than ten billion euros per year. This chapter takes its cue precisely from the central importance of unmanned aerial vehicles, commonly known as drones, and tries to sum up their future. Indeed, despite the enormous success of drones over the decades, many are the criticisms related to unmanned aircraft on board. The first section aims at outlining an historical framework, while the second section analyses the impressive role played by drones within the counterterrorism policy of the US presidency, with particular regard to Barack Obama's mandate. The third section applies a SWOT type analysis to UAVs, taking into account both their advantages and disadvantages. Finally, the fourth section investigates the human factor problem within UAVs, which was inspired by Dr. Giovanni Miranda's PhD thesis.

INTRODUCTION

During the first half of last January, the General Secretariat of the Council of the European Union produced an estimate that, by the year 2035, the drones' sector will create up to 100,000 new jobs and, in addition, will have an economic impact

DOI: 10.4018/978-1-7998-7126-2.ch012

of more than ten billion euros per year. The use of these technologies, which have been constantly implemented and perfected over time, has seen strong growth, especially in recent decades. To date, new drones are designed and continuously improved, allowing unprecedented results to be achieved. And it is precisely from the central importance of Unmanned aerial vehicles, commonly known as drones, and the many research I have come across that this chapter takes its cue and considers both their strengths and weaknesses, trying to sum up the future and the ambitions of drones. Indeed, despite the enormous success of drones over the decades, and in spite of their predominant presence among the foreign policy agenda of the last three American Presidencies, many are the criticisms of unmanned aircraft on board, and many are the challenges and dilemmas related to their use. The first paragraph aims at outlining an historical framework through which, then, a more detailed analysis will be carried out, and will briefly describe their typologies and functions. The second paragraph analyses the impressive role played by drones within the counter-terrorism policy of the US presidency, with particular regard to Barack Obama's mandate. Therefore, the use of drones within the broad sphere of Global War on Terrorism launched by George W. Bush following the al-Qaeda attack on the Twin Towers on September 11th, 2001 and pursued by Obama, particularly in the territories of Yemen, Pakistan, Somalia and Afghanistan, is investigated. The third paragraph applies a SWOT (Strengths, Weaknesses, Opportunities and Threats) type analysis mechanism to remotely piloted aircraft, taking into account the strengths and advantages of this technology and then exploring the disadvantages it implies. In the fourth and last paragraph, an investigation was proposed into the Human Factor problem within remote-piloted aircraft, which was inspired by Dr. Giovanni Miranda's doctoral thesis.

1. DEFINITION AND HISTORICAL ANALYSIS OF UAVS

According to the Cambridge Dictionary, the acronym UAV, Unmanned Aerial Vehicle, indicates aircraft without crew on board and the term "drone" is the most widely used synonym. Specifically, these aircraft use aerodynamic forces to help lift the vehicle, can fly autonomously or be piloted at a distance, can be expendable or recoverable and, in addition, can carry a lethal or non-lethal load, as The Free Dictionary states. Other expressions, such as robot planes, Remotely Piloted Vehicles (RPVs) and Remotely Piloted Aircrafts (RPAs) are also used to identify such aircraft (Cavoukian, 2012, p. 3).

Therefore, drones have the peculiarity and the advantage of not carrying a human operator inside them (Hassanalian & Abdelkefi, 2017, p. 99), and, according to the Office of the Secretary of Defense of the United States of America (2005), this characteristic makes this technology suitable to carry out those actions defined as dull, dirty and dangerous. In fact, the drones are respectively able to carry out flights lasting thirty hours or more without any failure or fatigue, as happens instead for the pilot; they can also cross contaminated places (such as, for example, clouds after the detonation of a bomb) managing to collect radioactive samples; finally, although the reconnaissance actions have always been dangerous, being normally conducted in hostile territories, the drones do not put the pilot's life at risk, as they do not actually possess one (OSD, 2005, p. 2).

The classifications of Unmanned Aerial Vehicles carried out by academics, scholars and international organizations are numerous, and they differ according to the parameters used (e.g., application and composition).

As far as the application of remotely piloted aircraft is concerned, it can be said that it has undergone considerable variations over the decades. If the Austrians, during the First Italian War of Independence in 1849, used the first "prototypes" of drones solely to discourage the Venetian resistance (Mckenna, 2016, p. 355), the multiple functions of today's Reaper drone - or MQ-9 - extend far beyond (Read, 2020), showing the incessant progress of technology over the centuries.

Professor Shaw, professor of human geography at the University of Glasgow, reports in an online article (2014) on the evolution of the American drone war, highlighting the passage of the war through five distinct phases, which can be seen in Figure 1.

While at the beginning of the twentieth century UAVs were used, in military function, essentially as targets, dummies with which to carry out training - as is the case of the automatic airplane Hewitt-Sperry and Kettering Bug -, in the period between the two World Wars they, including the B-17 bomber, employed in Operation Aphrodite, were used to transport explosives and report orders through enemy lines.

The third phase refers to the Cold War, in particular the period following the Cuban Missile Crisis, which took place in October 1962. In this phase the reigning function is that of "surveillance drone"; in this sense, in fact, they were used to collect photographic material in areas considered dangerous or with denied access. The example par excellence of surveillance drone is represented by Lightning Bug.

Predator and Reaper dominate the fourth phase, catalogued by the author as the "hunter-killer drone" phase. In fact, these two drones, protagonists of the historical period that began with the terrorist episode of September 11, 2001, also known as the War on Terrorism, are characterized by their massive use in the manhunt of this period, mainly in the hot areas of Afghanistan, Pakistan, Yemen and Somalia.

The fifth and final phase, whose historical period coincides with that of the fourth phase, brings together all those drones (such as, for example, the Global Hawk) which are used by the police forces to control and guard cities, borders and regions (Shaw, 2014).

Figure 1. The evolutionary phases of the American Drone War

EVOLUTION OF US DRONE WARFARE	Phase 1 **The Target Drone**	Phase 2 **The Flying Bomb Drone**	Phase 3 **The Surveillance Drone**	Phase 4 **The Hunter-Killer Drone**	Phase 5 **The Police Drone**
Time period	Early twentieth century	Interwar period	Cold War, particularly after the Cuban Missile Crisis	Post-September 11, 2001	Post-September 11, 2001
Institutions	U.S. Army	U.S. Air Force	U.S. Air Force and Strategic Command	U.S. Air Force, CIA and JSOC	Police forces, Customs and Border Patrol
Military logic	Drones were used as practice dummies	Drones were used to deliver ordinance across enemy lines	Drones were used to photograph denied or dangerous areas	Drones were used in a military "manhunt" during the war on terror	Drones are being used by police forces in the U.S and Europe
Key Geographies	Developed in UK and US military shooting ranges	Trialed across English Channel	Drones were used across North Vietnam, Cuba, China	Afghanistan (AF), Pakistan, Yemen and Somalia (CIA and JSOC)	Cities in the global North, used by hobbyists and criminals
Spatial Logic	-	Cross the Battlefield, Bomb the Nation State	Surveil the Battlespace, Capture photos	Hunt the Battlespace for dangerous individuals	Swarm the Street, protect VIP buildings
Iconic drones	Hewitt-Sperry Automatic Airplane, Kettering Torpedo	Glide Bomb, Modified B-17 bomber (Aphrodite)	Various Firebee drones, also known as Lightning Bugs	Predator, and later Reaper drone, also the hand-held Raven	Various quadcopter drones

However, Shaw, in conclusion, specifies that the fifth and final phase, that of the "police" drone, has just landed, and therefore bequeaths to readers a, so to speak, "open" ending about a possible future in which drones are monitoring entire cities, perhaps even armed, reporting an incident dating back to 2015, when a drone carrying an active substance landed on the office of the Japanese Prime Minister.

2. THE ROLE OF DRONES WITHIN PRESIDENT OBAMA'S COUNTER-TERRORISM POLICY

The United States of America has always represented a fundamental element in the understanding of drones and their use, which, particularly during the last three presidencies, has undergone considerable growth. When Barack Obama became president in 2008, the political situation he faced was undoubtedly a delicate one: not only was a new economic crisis looming on the horizon - and one that, moreover, had its epicenter in the United States of America - but the threat of new transnational players on the international scene was becoming increasingly tangible, especially within a country that only a few years earlier, on September 11, 2001, had suffered four major terrorist attacks, perpetrated by the Islamist terrorist movement al-Qaida, targeting the Twin Towers of the World Trade Center, the Pentagon, and a scheduled flight later crashed near Shanksville, Pennsylvania.

Obama's predecessor, George Walker Bush, who had become president only a few months earlier, on 20 January 2001, in response to these attacks, decided to adopt what would be known as the Global War on Terror. It was during his speech on September 20, 2001, which preceded a joint session of Congress on the United States' response to these terrorist attacks, that Bush Jr. pronounced the words quoted here (2001):

<< *[...] Our enemy is a radical network of terrorists, as well as any government that encourages them. Our war on terror begins with Al Qaeda but does not end here. It will not end until every terrorist group of global reach is found, stopped and defeated.*>>

These words were followed by action. Over the years, Bush took numerous actions in the name of the War on terror, from "Operation Enduring Freedom" (OEF) in Afghanistan, the Philippines and the Horn of Africa, to the Second Gulf War in 2003, which began with the US invasion of Iraq, to the execution of Saddam Hussein on 30 December 2006 in Baghdad (History.com Editors, 2019).

In particular, Bush resumed and expanded the policy of targeted killings, i.e. the targeted killings that his predecessor, Bill Clinton, had used a few years earlier, when he issued three Memoranda of Understanding authorizing the CIA to annihilate Bin Laden and the key elements among the latter's lieutenants. Bush, just six days after the terrorist attack, signed a Memorandum of Notification which, first, authorized the CIA to eliminate about two dozen or so leaders who appeared on a "list of high-value targets" and, second, allowed the intelligence agency to act without further presidential approval (Zenko, 2011).

A feature of Bush's War on terror policy is the implementation of the strategy of targeted assassinations, especially through the use of drones. In particular, convinced that a higher level of surveillance would have prevented the attacks of 11 September, and also terrified by the possibility of omitting some signs of potential terrorist threats, the American authorities began to link multiple ordinary crimes to terrorist acts (Priest & Arkin, 2011, p. xxi).

Specifically, on September 18, 2001, the American Congress passed a law, the Authorization for Use of Military Force (AUMF), which in section two, paragraph a, stated that:

<< *[...] The President is authorized to use all necessary and appropriate forces against those nations, organizations or persons whom he considers having planned, authorized, committed or assisted in the terrorist attacks that took place on September 11, 2001, or hosted such organizations or persons, in order to prevent the occurrence of other future acts of international terrorism against the United States of America by those nations, organizations or persons.*>>

With the entry into force of the AUMF Act, on the one hand, a greater amount of power was given to the presidential figure, who from then on had the power to decide, first and foremost, which persons should be held responsible for the attacks of 11 September and, secondly, how to deal with the problem of terrorism; on the other hand, the CIA was given greater flexibility from an operational point of view which, in the words of an expression quoted in chapter 1, was translated into the famous phrase of Cofer Black, Coordinator for counter-terrorism policies of the Bush administration, in which it is stated that, after 9/11, "the gloves come off".

The passing of the baton from George Bush Jr. to Barack Obama was, in some ways, experienced as a watershed, especially because of the difference of opinion of the two presidents about the 2003 Iraq War, which Obama strongly opposed (Zeleny, 2007).

However, once he became President, following his victory in the elections on November 4, 2008, Obama's policy proceeded in continuity with that of his Republican predecessor. In the words of David Brooks, commentator for the New York Times, when Obama was a Senator, he was not accountable for the leadership of the country and the implications that such a figure entails; instead, once he became President, he was reduced by the dynamics of a world in which there is a long-term war against al-Qaeda and, above all, a world in which drones can effectively eliminate terrorists, reducing the civilian victims involved in such actions. In simple terms, "brutal action abroad saves lives at home" (Brooks, 2013).

Therefore, as soon as he became President, Obama was rather reluctant about the use of signature strikes, i.e. targeted murders, in particular directed towards MAM, an acronym for military-age evil, individuals considered guilty of criminal and terrorist acts on the basis of their demography and not on concrete evidence (Zenko, July 16, 2012). The political scientist Micah Zenko, examining the book "Kill or Capture: the War on Terror and the soul of the Obama Presidency" by Daniel Klaidman, states that Obama encountered, on the former, a certain difficulty in understanding the concept of signature strikes, different from the so-called high-value individual strikes, in which the terrorist leader was identified before the launch of the missiles.

However, when in October 2009, Leon Panetta, who had been appointed director of the CIA a few months earlier by Obama, presented Obama - at the Situation Room meeting at the White House - with the list containing the ten subjects chosen as terrorist targets to be eliminated, the results obtained exceeded all expectations. In fact, not only did Panetta obtain approval for all ten subjects, but also additional Predator drones, additional resources for the CIA intelligence activity and, finally, the authority to pursue the so-called Pakistani "target boxes", that is, the areas within the tribal regions where the CIA could operate (Zenko, July 2, 2012).

According to an article written by journalist Michael Hastings, when Obama became President, he inherited two drone programs. The first, sponsored by the Pentagon, focused primarily on providing reconnaissance and air strikes to protect American ground troops, was a program that operated almost always in the open, and was based on more than a dozen military centers located around the globe, centers within which lawyers were present, with the task of verifying that any action was taken in accordance with the Geneva Conventions. The second program, the CIA program, differed considerably from the first: first, the CIA operated secretly within this program; second, although the CIA lawyers had to sign off on the drones' attacks, the process remained confidential, and the supervision to which it was subjected was considerably less than that provided by the military; last but not least, the CIA conducted such attacks in places where the United States of America was not officially in a state of war, such as Yemen, Somalia and Pakistan (Hastings, 2012).

The graph shown in Figure 11 (Woods & Ross, 2012) shows the analysis carried out by two journalists of The Bureau of Investigative Journalism, which examines the attacks carried out by drones during Obama's first mandate (2008-2012). Specifically, the graph shows that, as far as Libya is concerned, Obama used drones during 2011, since the first Libyan civil war took place between February and October of that year. Moreover, during the first four years of the Presidency, the attacks operated in Iraq by remotely piloted aircraft were concentrated in 2008 and 2009, and in particular were carried out by the Reaper drones of the USAF (United States Air Force). Yemen is a territory in which attacks using drones were carried out from 2011 onwards, the year in which the Arab Spring and the Yemeni Revolution took place. In conclusion, Pakistan and Afghanistan represent a "hot" territory where the Obamian warfare drone has been applied on numerous occasions during the four-year period 2008-2012.

Figure 2. "Drone strikes" during Obama's first term in office

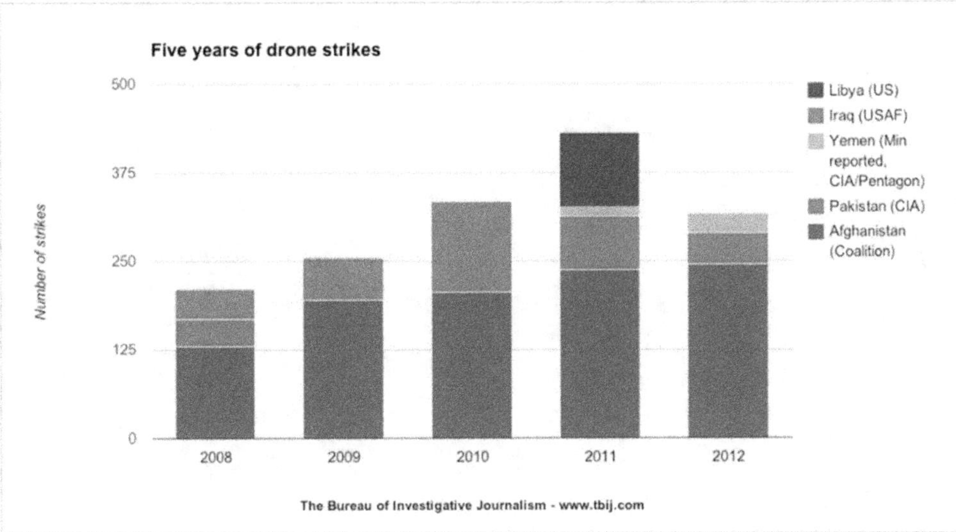

2.1 Yemen (2002-2020)

As far as the state of Yemen is concerned, with the exception of a single strike drone episode that took place on Sunday, November 3, 2002 under the presidential administration of George W. Bush, which had 6 men as victims, it is possible to state that a real program of using remotely piloted aircraft did not start until 2009, when Obama became president, when it was called "playbook drone" (Bergen et al., 2002).

Although the count of drone and air strikes in Yemen is complicated for many reasons, as some scholars have pointed out (Bergen & Rowland, 2012), Figure 13, taken from a research project started eighteen years ago and still ongoing (Bergen et al., 2002), aims to illustrate the distribution of drone attacks in Yemen between 2002 and 2020, and therefore takes into account three different US presidencies - Bush Jr., Obama and Trump. During the period of interest of my research, which includes the years of Obama's presidency, from 2009 to 2016, it is possible to notice a surge in the use of remote-piloting aircraft.

Indeed, at the beginning of 2009, al-Qaeda militants, expelled from Saudi Arabia thanks to Barack Obama's anti-terrorist policy, together with Yemeni militants, declared the formation of an al-Qaeda affiliated group in the Arabian Peninsula, better known as AQAP (Al-Qaeda in the Arabian Peninsula). Between 2009 and 2010, when an AQAP affiliate, Umar Farouk Abdulmutallab, attempted to bomb a plane from Amsterdam to Detroit, Michigan on 25 December 2009, President Obama increased the attacks (by air and drone) in Yemen (Fischer, 2014, p. 33).

Figure 3. Attacks conducted by drones in Yemen between 2002 and 2020

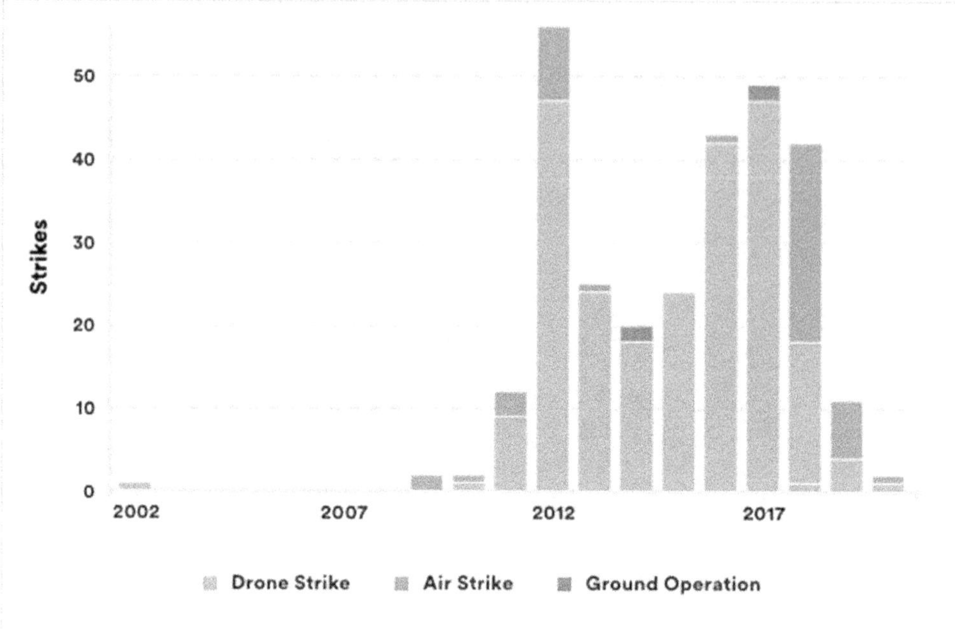

When AQAP conducted numerous attacks on local security forces and intervened with territorial claims in the context of the 2011 Arab Spring demonstrations in Yemen, the United States further increased both air and drone attacks in Yemen (Bergen et al., 2002).

On 30 September 2011, the assassination of the "Professor", Anwar al-Awlaki - a well-known American terror propagandist who had fled the United States in 2002 after being accused of suspected terrorist activity - took place through a drone attack (Lister, 2011). This event is the first reported case of the deliberate elimination of an American citizen since the War of Secession (1861-65). Because of this, together with the fact that the assassination of al-Awlaki took place without trial, a strong criticism of Obama was made by the Republicans, who harshly criticized the lack of official evidence regarding the terrorist's membership of the al-Qaeda network (Orban, 2014, pp. 7-8).

After 2011, the year in which there were 3 air attacks and 9 attacks using drones, the attacks increased considerably, reaching a peak in 2012, with 9 air attacks and 47 attacks using drones. In the following years the attacks decreased slowly, especially in 2014, although in 2015 a new surge in the frequency curve of drone-based attacks began, due mainly to the extensive progress within the country by the Yemeni Houthi rebels, who renewed their conflict with the Americans in 2015. The latter, also due to the renewed attacks by AQAP, began to restart their plans for the war (Bergen et al., 2002).

2.2 Pakistan (2011)

One of Obama's greatest successes was the successful raid in Pakistan during which Osama bin Laden was assassinated on the night of 1 May 2011 in Abbottabad. In fact, although the United States launched its first attack operation in Pakistan - using drones - as early as 19 June 2004 under George W. Bush's administration, it was under Barack Obama's presidency that actions in this state increased significantly.

In particular, the number of attacks that President Bush conducted on Pakistani territory represents only a small fraction compared to those conducted by his successor. Given in hand, while Bush ordered a total of 48 attacks in Pakistan during his presidency, Obama ordered 353. Some researchers who were part of a project concerning drone-based attacks in Pakistan report the statistics that can be seen in Figure 4 (Bergen et al., 2004).

Figure 4. Attacks conducted by drones in Pakistan sorted by year and administration

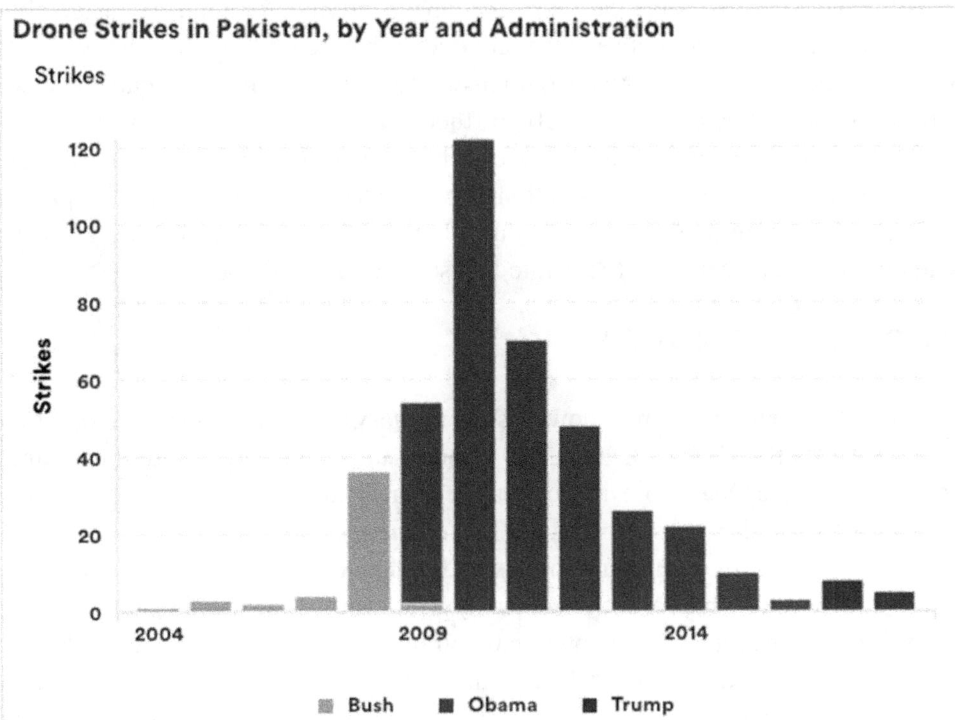

As the graph in Figure 4 shows, Obama significantly increased the number of drone attacks since 2009, when Bush executed two attacks against Obama's 52. A real peak can be seen in 2010, with a total of 122 actions, while since 2011 the numbers have fallen again, from 70 in 2011, to 48 in 2012, to 26 in 2013 and 22, 10 and 3 in 2014, 2015 and 2016 respectively.

The American Drone War in Pakistan took place in the so-called FATA (Federally Administered Tribal Areas), comprising the portion of Pakistani territory between the Afghan border and the North-West Frontier Province. Since its inception, this war has had numerous assassinations of both al-Qaeda and Taliban leaders. In this regard, the research project mentioned above (Bergen et al., 2004) has analyzed the change of direction that the three US administrations have pursued over the years with respect to the target groups of their campaign. In particular, if al-Qaeda has been the main target of the Bush Jr. administration. (for a total of 18 attacks out of a total of 48, i.e. 38%), for Obama the movement is in fourth place on the list of targeted killings; in fact, while Obama's target par excellence was represented by targets whose membership was not transparent and is indicated with Unclear (153 of the total 353 attacks, i.e. 43%), followed by the Taliban (114 of the total 353 attacks, i.e. 32%) and the Haqqani Network (34 of the total 353 attacks, i.e. 10%), with al-Qaeda having even fewer attacks (33 of the total 353, i.e. 9%).

An article in the US magazine The New York Times (Mazzetti & Sanger, 2010) pointed out that, over the years, the Obama administration has considerably extended what is called a covert war, conducted by the CIA intelligence services within the Pakistani state, which would justify the increase and consequent peak of attacks conducted by drones in 2010. Furthermore, if President Bush had secretly decided, in his last year in office, to extend the drone program, Obama has not only continued along the same line of action as his predecessor, but has also simplified its use, allowing the CIA director, Leon E. Panetta, to authorize numerous attacks without prior notification to the White House (Cloud, 2010).

2.3 Somalia (2004-2019)

Although the United States has shown lasting military involvement in Somalia, dating back to the Battle of Mogadishu in 1993, fought under the Clinton administration, in the years of the Obama presidency interventions in this region were substantially reduced compared to those conducted in Yemen and Pakistan in the same years. In fact, the total sum of military actions undertaken by Obama in Somali territory amounts to 34, 3 of which are represented by ground operations, while the remaining 31 by air or drone attacks, as shown in the graph in Figure 5 (Bergen et al, 2003).

As Stig Jarle Hansen says (2019), drones in Somalia are a valid weapon without doubt, but they are not a "game changer". In fact, although in January 2007 the

Figure 5. Drone attacks conducted in Somalia through the use of drones sorted by year and administration

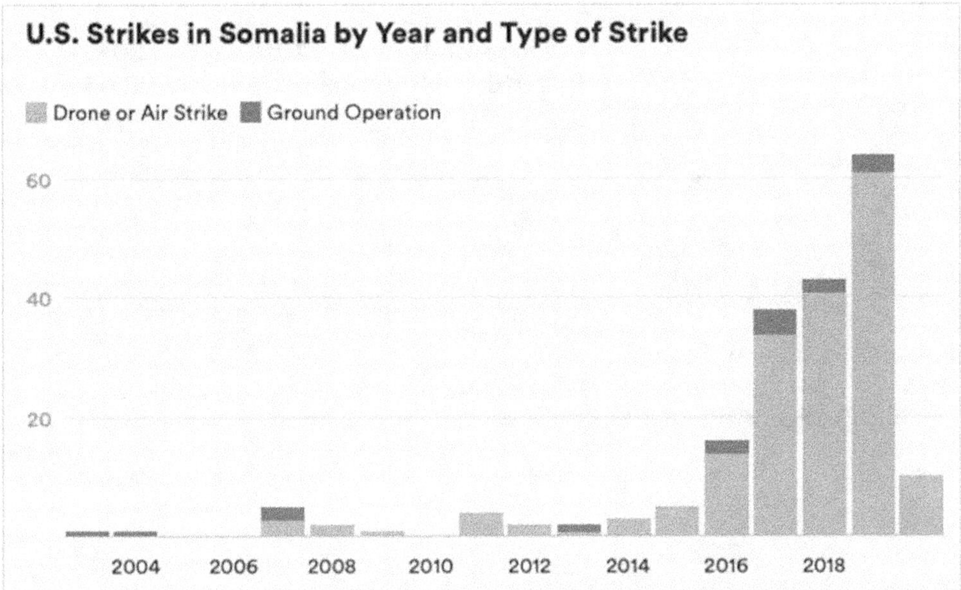

United States used a Predator drone to locate the al-Qaeda operational convoy led by Abu Taha Al Sudani, until June 2011 drones were only used in combination with cruise missiles, gunners and attack helicopters.

In any case, according to Hansen (2019), the intensification of the attacks conducted by the drones in Somalia has not, in fact, succeeded in promoting a strategic victory against the Somali al-Qaeda cell, i.e., the terrorist group of Al-Shabaab, active in this territory since 2006. Moreover, the US General in charge of the Commando in Africa himself endorsed this aspect before the Senate Armed Services Committee in February 2019 (Agency France-Presse, 2019).

2.4 Iraq (2001-2003) and Afghanistan (2015-2016)

Last but not least, the use of drones in Iraq and Afghanistan has, as for the states analyzed in the previous paragraphs, a long history. As far as the first of the two states is concerned, already during the first Gulf War the United States started to fly drones over Iraq. In addition, in the late 1990s, then-President Saddam Hussein launched an Iraqi drone development program - which the US would later use to justify the invasion of Iraq in 2003. It was during the latter that the Americans began to make increasing use of drones (Gettinger, 2020). Since the origins of George

Figure 6. US drone-operated attacks and other types of air strikes in Afghanistan

Afghanistan: Bureau data on US drone strikes and other airstrikes

	December 2016	2016	2015-2016
US strikes	8	1071	1306-1307
Total people reported killed	24-26	1389-1597	2371-3031
Civilians reported killed	0	65-105	125-182
Children reported killed	0	3-7	6-23
Total people reported injured	12	196-243	338-390

W. Bush's War on Terror, to be found in the terrorist attacks of 11 September 2001, Predator and Reaper drones have been used in both Afghanistan and Iraq. However, as far as Iraq was concerned, the use of drones was, to tell the truth, strictly limited, since once he became President, Obama announced, on 31 August of ten years ago, a clear change of direction compared to his predecessor: it was time to end the combat mission in Iraq, which had begun during 2003. On December 21, 2011, Obama went to Fort Bragg to emphasize the withdrawal of the last American soldiers from Iraq, officially ending the war in this state (The White House, 2009). Also, in Afghanistan, as in Iraq, armed drones have been used for several years. In fact, US involvement in the Afghan region has increased since 2001, when the war in Afghanistan broke out and is still ongoing. US use of drones in this area is undoubtedly reduced compared to their use in other areas, such as Pakistan and Yemen, but it is not non-existent. Figure 6 (Purkiss & Serle, 2017) shows Obama's numerical involvement in this region until December 2016.

It has been repeatedly pointed out that the use of drones in the Afghan conflict has not been visible, unlike in the FATA areas of Pakistan or in Yemen and Somalia (Clark, 2017). The same authors cited above specify, in their research, that acquiring precise data about the air attacks conducted by the United States in Afghanistan has led to data processing which, due to the war still in progress, is not exhaustive. American warplanes launched 1,337 missiles on Afghanistan during 2016, an increase of 40% compared to 2015, according to data released by the US Air Force. This increase is justified by Obama's decision, taken in June 2015, to give US commanders more leeway to hit the Taliban and provide aid to the Afghan army in order to prevent strategic cities from falling into the hands of insurgents (Purkiss & Serle, 2017).

In fact, this decision represented a departure from the current US rules for Afghanistan, which imposed limitations on the ability of US forces to strike at insurgents. For example, before Obama's decision, US soldiers could only intervene against the Taliban "in extremis", i.e., in situations where there was a concrete need for their intervention to avoid significant setbacks by Afghan soldiers.

Thanks to this new Obama policy, however, the US military were, from that moment on, authorized to accompany the Afghans in key moments of their offensive campaign against the Taliban (Stewart, 2016).

Moreover, while from 20 December 2001 until December 2014 there was ISAF (International Security Assistance Force), the NATO (North Atlantic Treaty Organization) mission authorized by the United Nations (UN), which aimed to provide support to the Afghan government in the war against the Taliban and al-Qaeda, when the latter expired on 1 January 2015, the NATO-led mission known as RSM (Operation Resolute Support Mission) took its place. The objective of this mission follows that of the previous one, namely, to prepare, assist and advise the Afghan police and army. In support of the mission, the United States has considerable firepower (Serle & Sargand (2015): in fact, the Air Force manages not only the F-16 attack (aircraft stationed at Bagram air base near Kabul), but also the Predator and Reaper drones based in Kandahar and Jalalabad; there are also Apache attack helicopters available on the territory; finally, the CIA still operates with drones in Afghanistan, while the US Air Force continues to fly AC-130 guns.

2.5 Advantages and Criticisms

In the entire history of the United States, says Greg Miller (2011), no other president has ever relied as much on secret killings to keep the country and its objectives safe as Barack Obama.

However, Obama's extensive use of remotely piloted aircraft has not been without criticism. In fact, drones constantly question legal regulations and, in the case of the United States, their use - as far as territories with which they themselves were not technically at war - was justified in the name of the "War on Terror" (Sabbagh, 2019). The criticism of the policy of using Obamian drones varies considerably, depending on the thematic area considered. In fact, an academic like Rosa Brooks (2013), who would later make proposals for some improvements to the war on drones, said (Brooks, 2012):

<<Drones have seriously become the Obama administration's weapon of choice in the fight against terrorism. What I think hasn't been done enough is to take a step back and ask ourselves, 'Are we creating more terrorists than we are killing? Are we promoting militarism and extremism in the very places where we are trying to eliminate it?'. Much of what is happening with the drone attacks is still shrouded in secrecy. It is very difficult to assess from the outside how serious the threat posed by those affected is.>>

On the other hand, the object of criticism was the lack of transparency with regard to civilian victims that the drone attacks caused. In fact, according to The Bureau of Investigative Journalism (Purkiss & Serle, 2017), while US estimates of civilian casualties are between 64 and 116 individuals, the number would be quite different, ranging from 380 to 801, which is about six times higher.

The 25th director of the CIA, John Brennan, expressed this view in a public speech in 2012 (Wilson Center, 2012), saying that not only is Obama's policy of targeted homicides legal - since the US Constitution gives the president the power to protect the nation from any imminent terrorist attack with AUMF authorization - but also ethical, since such targeted executions comply with the principle of necessity - that is, the fact that the target is given a specific military value, as in the case of individuals belonging to al-Qaeda or forces associated with the latter - as well as the distinction between them - since only military and non-civil targets can be intentionally targeted - and proportionality - since remotely piloted aircraft target a single terrorist, or at least a small number, thus managing to minimize the risk of collateral damage to the civilian population.

In truth, some sources (Woods, 2011) categorically deny what Brennan said, claiming, on the contrary, that a detailed examination of the secret attacks conducted by the CIA by drones in Pakistan since August 2010 has brought to light at least a dozen individual attacks, which resulted in the deaths of forty-five or more civilians, information reported earlier this year also by the Financial Times newspaper (Warrell, 2020). Finally, as a "justification" for this, Council on Foreign Relations (Zenko, 2012) reports that US officials rely on a purely logical reasoning that people within an area of terrorist activity probably do not have good intentions.

Moreover, while some critical voices (Feroz, 2020) would like the killings of civilians to be thoroughly investigated and cases brought before the International Criminal Court, others (Currier & Maass, 2015) denounce the problem of the "tyranny of distance", i.e., the inability of the US military to carry out a full-time surveillance system against those identified as targets in the areas of Yemen and the Horn of Africa.

On the other hand, there has been no lack of observations over the years about the many advantages of using drones. In fact, it has been stressed (Mir, 2018) that the political program on the use of drones has enormous potential for inflicting damage on armed target groups in order to undermine their organization and their ability to mount attacks against the United States of America.

Moreover, the article by journalist Michael Hastings (2012) highlights the convenience - defined with the attribute "lethal" - of drones which, thanks to their low cost and their intrinsic advantage, i.e., the fact that their use allows the realization of targeted assassinations at a distance, have become, over the years, a "must-have item", an essential element for any advanced military power. In fact, even though the drones used in Iraq and Afghanistan have a cost of thirteen million dollars apiece, the global market for remotely piloted aircraft was already travelling at a speed of six billion dollars a year in 2012, when the article was written. To date, estimates have been made (Business Insider Intelligence, 2020) that by 2025 drones' shipments will total $63.6 billions; moreover, by 2028 the US is expected to purchase more light surveillance drones than the rest of the world. In short, the so-called "disposition matrix" policy, the Orwellian euphemism referring to Obama's policy of targeted assassinations (Greenwald, 2012), has been pursued and institutionalized during the years of the Obama presidency. All this has undoubtedly led to some consequences. These include the emergence of strongly anti-American feelings in Pakistan, as well as a progressive deterioration in diplomatic relations between these two countries. Moreover, the drone war, going on with the years, has extended within the Pakistani state, moving from the less populated tribal areas to the more populated ones, such as Peshawar, contributing to the fomentation of the inhabitants, which, in turn, undermines not only Pakistan's sovereignty and government legitimacy, but also its ability to build a unitary and solid Pakistani state. Nevertheless, there is no lack of criticism about a "tactical disconnection" between the units responsible for the actions conducted by drones, i.e. the military, the JSOC (Joint Special Operations Command), and the paramilitary, the CIA, which have been in conflict several times over the years. In conclusion, numerous academics and diplomats (Vainauskas, 2014) have discussed the joint existence within the war of drones, on the one hand, of tactical successes in the short term and, on the other, of strategic costs in the long term.

The next chapter will propose an analysis of the costs and benefits of the use of drones, but in a global sense, i.e., not limited to the United States of America. After all, as a former CIA analyst and Barack Obama's counter-terrorism policy advisor said (Miller et al., 2012):

<<*The problem with drones is that they are like your lawnmowers. You have to mow the lawn all the time. The moment you stop mowing, the grass will grow back.*>>

3. SWOT ANALYSIS APPLIED TO THE USE OF DRONES

The third paragraph of this chapter makes use of SWOT analysis (Vocabolario Treccani), which is a tool used in the formulation of management and strategic planning in order to shed light on certain aspects with reference to an organization, a company or, in the case examined by this thesis research, a production within a specific market, such as that of drones.

In particular, the acronym SWOT identifies the elements to be analyzed: Strengths, Weakness, Opportunities and, finally, Threats of the object considered. These four elements are arranged within a matrix, like the one shown in Figure 7 (Shewan, 2020).

Specifically, strengths and opportunities represent the positive elements detected by the analysis, while weaknesses and threats are the negative ones. At the same time, strengths and weaknesses fall within the framework of an internal study of the object being analyzed; on the other hand, opportunities and threats consider elements external to the object itself.

Figure 7. Example of SWOT analysis matrix

3.1 Strengths of Unmanned Aerial Vehicles

The last few decades have shown that remotely piloted aircraft can be used in the most diverse situations, whether civil or military. In fact, drones can count numerous supporters among their ranks, starting with the famous US e-commerce company, Amazon.com, which, after obtaining the necessary authorizations from the US Federal Aviation Administration (FAA), announced the project - which later became a reality - that foresees the use of drones to make shipments and deliveries of online purchases (Lee, 2019).

Moreover, as has been pointed out in the first chapter, the drones are also successfully used for further civil purposes, including environmental protection, whale watching, coral reef protection and, in the journalistic field, the collection of information about events in real time (Lee-Morrison, 2014).

The strengths of drones in the military sector are emphasized equally, if not more so, on several occasions. In this regard, it should be underlined first of all that, if on the one hand, many states are making increasing use of remotely piloted aircraft (Hambling, 2019), on the other hand (Orban, 2015), even at a global level, estimates are particularly encouraging in this sector, since it is expected that, given the expansion of the fleets of remotely piloted aircraft by an ever-increasing number of military powers, by 2030 UAVs could make up 50% of their total aircraft.

The drones make it possible to achieve various objectives at an operational level (Hoffman & Kalikow, 2013), the first of which is to limit civil losses as much as possible, a goal that cannot be achieved if a war is started using ground troops, where this occurs to a greater extent (Bowden, 2013). With respect to this point, in 2009, the then director of the CIA, Leon Panetta, affirmed with conviction (Shachtman, 2009) that, if, on the one hand, the remotely piloted aircraft proved to be more than effective in eliminating the suspect militants in Pakistan, above all, they represented the only truly effective weapon (literally, "the only game in town") to face al-Qaeda.

Indeed, as reported in the first chapter, unmanned aircraft have an undoubted advantage when compared to manned aircraft, as the fact that the former are certainly better suited to carrying out activities identifiable by the presence of at least one of the three attributes of the "three D's", dull, dirty and dangerous, such as routine surveillance activities, operations carried out at high altitudes or subject to contamination by unexploded mines (Alley, 2013, p. 4). In this sense, some sentences by Gordon Johnson, member of the Joint Forces Command of the Pentagon, encapsulate the scope of this discourse (Weiner, 2005):

<<*They are not hungry. They are not afraid. They do not forget the orders that are given to them. They don't care if the guy next to them has just been killed. Will they do a better job than humans? Yes.*>>

So, drones are often painted as robots that are, so to speak, immune to the physiological needs that grip the human being, such as hunger and thirst. Furthermore, drones would be foreign to feelings such as boredom and fun, so it is assumed that they carry out the tasks entrusted to them without delay, delays and, above all, miscalculations or oversights. In addition to this, precisely because of their diversity from the human being, some drones can boast a practically perfect view, a characteristic that has earned them an analogy with the Gorgon, a mythical tripartite creature of ancient Greece whose eyes, which never closed, had the power to petrify and turn into stone anyone who laid their eyes on her (Nakashima & Whitlock, 2011).

Maurizio Di Loreto (2006), in his research thesis, talks about the unique ability to use UAVs, referring respectively to their techniques of persistence, penetration and proximity. The "persistence" of UAVs allows them to operate non-stop for surprisingly long periods of time - it is estimated (Lewis, 2012, pp. 296-297) that some drones are able to remain in flight without being refueled for up to thirty hours overall - allowing them to operate with a level of autonomy naturally superior to that of manned aircraft. This autonomy, in turn, favors the approach to the target, establishing a link of proximity with the latter, creating an ideal situation to penetrate its defenses (Nakashima & Whitlock, 2011).

3.2 Opportunities of UAVS

In addition to these distinctive features, Michael Lewis (2012, p. 296) says that the main reason why it was decided, in fact, to use drone technology is the desire to minimize human lives, which in collecting information and executing orders put their lives at risk. A further decisive factor in this respect is undoubtedly the relatively low cost of drones compared to modern combat aircraft. In fact, while the economic value of a single Predator drone is approximately $5 million, and the cost of a Reaper drone is approximately between $10 and $15 million, the cost of an F-22 fighter aircraft is approximately $412 million.

From the operative point of view, from the above-mentioned advantage inherent to the dilated flight capabilities of UAVs derives another one, that is that, just in virtue of the time spent in flight observing the target, the drone possesses an accuracy that does not leave room for the possibility that an "unproportioned", indiscriminate attack may occur (Lewis, 2012, p. 297). In spite of the fact that the following paragraph deals in depth with the legal issues related to the figure of UAVs, it is important to underline that those who support the work of these aircraft believe that they are responsible for precision and accuracy, elements which are more difficult to attribute to manned aircraft.

It follows that the sum total of the advantages offered by the drones makes them the weapon of choice of the United States of America to fight the global war on terrorism, which I discussed during the second chapter. In fact, Byman (2013) reviews the reasons why the UAVs are so reliable for Washington, explaining that "nobody does it better" (literally, "nobody does it better") than drones, thus referring to estimates of terrorists and al-Qaeda militants that the drones were able to eliminate during the years of the war on terror (which, in 2013, exceeded three thousand). Thus, the drones have unquestionably demonstrated their usefulness, both by eliminating some veteran leaders of the terrorist organization and some elements located further down the chain of command but still dangerous, such as recruiters and bomb makers. In addition to this, the drones would have been able to debilitate the communicative capabilities of the organization itself, since the tight control of the remotely piloted aircraft makes it difficult - if not practically impossible - for the various al-Qaeda components to meet each other.

Furthermore, Byman maintains that there are some war zones or unstable countries - such as Pakistan, Yemen and Somalia - which, by definition, are unstable, and therefore, arresting militants in these areas is highly dangerous and, even if an operation in this sense were to be successful, it would often prove inefficient, since the governments of these countries have limited control of the remote areas of the country.

In addition, drones offer a very good ploy to the United States, because politically they make a kill rather than capture approach 'acceptable', meaning physical elimination rather than capture (Byman, 2013), since President Obama, during his second day in office, signed an executive order for the final closure of the detention center at the Guantanamo Bay naval station in Cuba (Masters, 2011).

Above all, being able to observe its target for hours, the remotely piloted aircraft waits for the right moment to hit it and, in doing so, greatly reduces the likelihood that a civilian could be involved in the assassination. In relation to this point, Byman highlights the fact that the number of civilians killed as a result of an attack by UAVs is unreliable, since defining who is civilian rather than military is often a difficult task; in this regard, he reports the episode that saw both the Pakistani Taliban leader Baitullah Mehsud and the doctor of the latter, who, even if he was not really a soldier, was in any case a figure who was helping a terrorist leader. Moreover, he continues, most attacks are carried out in areas of the country that are so difficult to reach that it is almost impossible for independent sources of information to verify exactly who was killed.

Finally, one must consider that the numbers and statistics are often and willingly manipulated by both the Pakistani government and activist groups, and that among these numbers and statistics there are also those concerning the killings of civilians by drones. In any case, Byman reports that the Bureau of Investigative Journalism estimates that the ratio of civilian casualties to military casualties - which are one to three among them - is lower than it would be if the attacks were carried out in other forms (Lewis, 2012, p. 297).

A further advantage of remotely piloted aircraft lies in the fact that they have earned the support of foreign governments, such as the Pakistani and Yemeni governments, which have seen attacks by drones as a strong ally against common enemies. Over the years, this support has come to the surface in a more veiled way, as when the President of Pakistan in office until 2008, Pervez Musharraf, stated that in Pakistan there are things falling from the sky at all times, and even more explicitly, as evidenced by the statement of the President of Yemen, Abdu Rabbu Mansour Hadi, that drones are able to locate the target, which, knowing the target, leaves a margin of error that is zero (Lewis, 2012, p. 297). In conclusion, in those places where terrorists actively operate against the United States of America, the drones would have acted at a very low cost, without endangering the lives of US soldiers and, above all, with minimal and certainly minor civilian losses compared to those that other means would have caused.

After analyzing the numerous advantages that the use of remotely piloted aircraft entails, there comes the feeling of being, in some way, faced with a phenomenon that we could describe as "bizarre". In fact, the multiple strengths of the drones

Figure 8. The impossible trilemma within counterinsurgency's operations

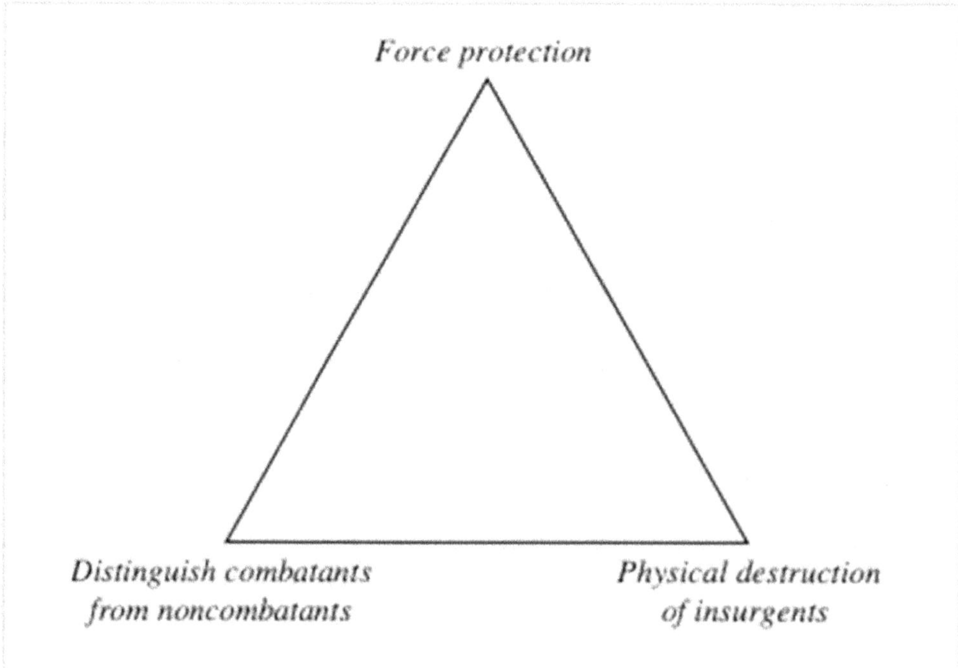

seem to dispel what Professor Lorenzo Zambernardi has defined as an "impossible trilemma" of counter-terrorism operations. According to Zambernardi (2010, p. 22), the common denominator of each counterinsurgent lies in the fact that it has three objectives: the first is the protection of its armed forces; the second is the distinction between enemy fighters and non-combatants; the last is the physical elimination of insurgents, as shown in Figure 8.

Applied to the context of a counterinsurgency situation, Zambernardi argues that it is impossible to achieve all three objectives listed at the same time, since to achieve two of them, it is necessary to "sacrifice" the third one. Yet drones seem to be the "perfect" weapon for a counterinsurgency campaign because they seem to be able to both eliminate terrorist units and protect their own forces, preventing soldiers from physically going to the battlefield and, at the same time, discriminating between those who fight and those who are civilians.

Overall, it is certainly not premature to say that drone technology will continue to play a major role in international markets. In addition to the reasons given above, it has been underlined that remotely piloted aircraft represent the ideal platform for multiple forms of information gathering and intelligence. In fact, it is precisely the combination of different types of intelligence that is one of the fundamental requirements for a precision attack to take effect. In particular, the forms of intelligence referred to are SIGINT, ELINT, COMINT and IMINT (Matějka, 2014, p. 31). SIGINT, acronym of Signals Intelligence, means the collection of information through the interception of signals emitted both by people and between machines. ELINT and COMINT, which indicate Electronic-signals Intelligence and Communications Intelligence are two SIGINT specializations, and respectively deal with the collection of information through the use of electronic sensors and the interception of extraneous voice or text communications. Finally, IMINT, which indicates the form of information collection Imagery Intelligence, finds information through the analysis of aerial or satellite photographs. If we consider the future prospects for the application of remotely piloted aircraft, it is possible to state that the market of drones, at least as far as their use at civil level is concerned, is destined to grow, both in Italy, where some sources state that, since 55% of the companies have registered a considerable increase in turnover in the drones sector. Over the next three years, we will see further growth, both at international level, as from 1 July 2020 the regulation of the European Union Aviation Safety Agency (Easa) will enter into force, which will initiate a process of liberalization of drones, which, in turn, will increase competition and consequently the sale of remotely piloted aircraft (Netti, 2020).

Figure 9 (Markets And Markets, 2018) shows that the military drone market is expected to grow to $26.8 billion by 2025. One of the determinants of this growth is attributed to the purchase of remotely piloted aircraft with various applications

Figure 9. Overview of the military drone market by geographic area, 2016-2025 (USD billion)

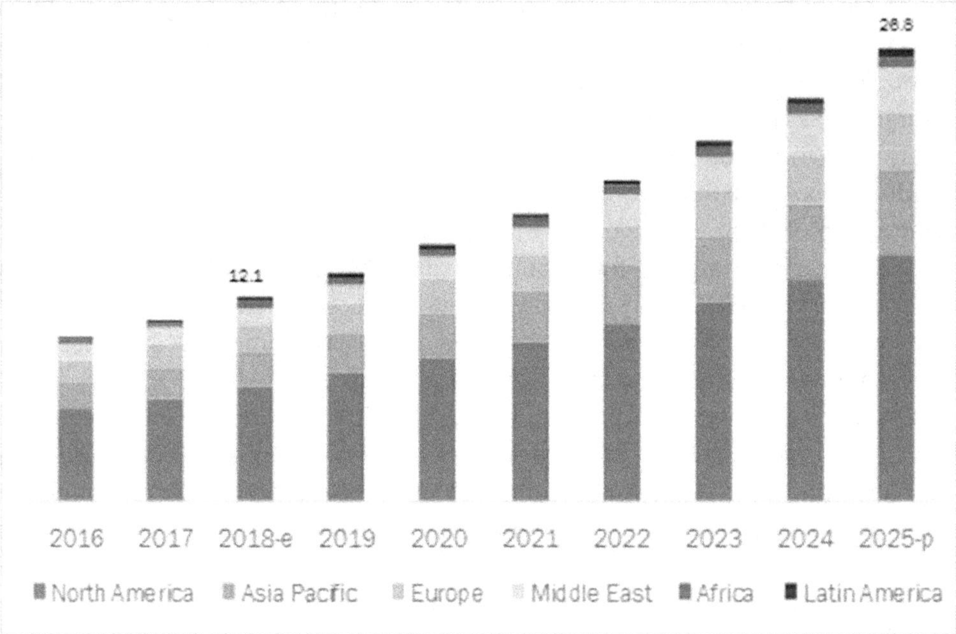

at a military level, such as the defense function, as well as monitoring, detection, mapping and combat operations.

Therefore, on the basis of what we have seen so far, it would seem that there is a rosy future full of expectations waiting for drones. However, the following paragraph will analyze the - numerous - criticisms that have been made of remotely piloted aircraft over the years. These observations highlight the disadvantageous sides of drones and paint their prospects for future use in tones that are far from reassuring and promising.

3.3 *Weaknesses* and *Threats* of UAVS

Although drones represent a valid alternative to conventional aerial missions – militarily speaking – over the years a huge number of criticisms and dilemmas, ethical and moral, have arisen among academics, scientists and journalists.

First of all, the fact that drones do not enjoy a very good reputation. Indeed, public opinion tends to associate the image of a remotely piloted aircraft like Predator with that of a modern Terminator. A further criticism in this sense is reported in an article in the Los Angeles Times (McManus, 2009), which quotes a significant phrase by David Kilcullen, counterinsurgency expert, who states that the use of drones, defined as robots, denotes an attitude that is cowardly and, at the same time, weak.

Secondly, it is necessary to consider the growing fear that it may become excessively "simple" to use technological instruments that by nature are both cheaper and safer for soldiers (from a physical point of view), thus implying a greater propensity to embark on conflicts, wars and missions, ending up putting civil lives at risk, thus transferring - and not completely eliminating - the danger to the latter (Sabbagh, 2019). This point has been studied on several occasions by various academics, including Sauer and Schörnig (2012), who claim that a high use of remotely piloted aircraft within an asymmetric conflict (i.e., a conflict between a state and a non-state actor) implies the risk of exacerbating the conflict, lowering the threshold of recourse to military means for the resolution of a conflict. In fact, estimates of drone attacks in states such as Afghanistan and Pakistan seem to suggest that drones are not as precise and discriminatory as one might think, not achieving the much-coveted effect of minimizing civilian casualties of war.

Taking up the subject of civilians killed by drones, it should first of all be stressed that, as can be easily deduced, the populations that are attacked by drones develop a negative attitude towards those who hit them. With respect to this point, there are several problems that emerge, including indifference to the deaths of civilians (Byrne, 2015, p. 6). Moreover, some journalists (Currier & Maass, 2015) reported that during some operations conducted by the United States in Yemen and Somalia, only a few Reaper drones were equipped with high-definition video, while most of the aircraft lacked the ability to collect recognition data; this problem of efficiency would explain why in these regions the attacks conducted by drones on military kill list targets were less successful - causing the death of the wrong people - than those conducted in Afghanistan and Iraq.

Furthermore, as an article in Foreign Affairs reports, there are several reasons to believe that not only are the drone attacks by the US not having the desired effect, i.e., the annihilation of the al-Qaeda terrorist organization, but that they are even strengthening it. These reasons include the fact that Washington was able to use drones to eliminate some of the leaders of the terrorist cell, but was unable to prevent their replacement, and al-Qaeda continued its proselytism and propaganda to recruit new terrorists. Last but not least, al-Qaeda has cunningly and on more than one occasion aimed to its own advantage at the killing of civilian victims by the Americans, painting them as "immoral bullies who care less about ordinary people than al-Qaeda" (Cronin, 2013).

Finally, a further concern with regard to the use of remotely piloted aircraft lies, according to some (Keene, 2015, pp. 21-22), in the fact that drone operators are led to develop a "PlayStation" mentality, as the physical distance from the battlefield would contribute to the ease with which you press the drone activation button. In this regard, the background to these concerns is the specter of a concept that Hannah Arendt has named "rule by Nobody", which we could define as "nobody's government", i.e., a reality in which it is not possible to locate a source of responsibility to blame for the fait accompli, and which Shaw and Akhter apply precisely to the use of drones in contemporary times (Shaw & Akhter, 2014, p. 229).

3.4 Legal Dilemmas

The drone, defined within the masterpiece *A Theory of the Drone* written by Grégoire Chamayou (2015), one of the most experienced in the field of UAVs, as "*unidentified violent object*" (p. 14) wholly redefines the main definition of war as it was originally conceived in Carl von Clausewitz's mind, namely "a duel on a large scale" (Clausewitz, 1832, book I, p. 13). Indeed, drone war is not a duel anymore; instead, it is more resemblant to a "hide-and-seek game" (Chamayou, 2015, pp. 33-34), in which there is a hunter advancing on a prey that flees or hides from him" (p. 33). And indeed, the paradoxical truth of the drone is that, by depriving the enemy of any possibility of participating directly in the hostile confrontation, the drone itself deprives itself of the ability to differentiate, which takes place at the very moment of combat (pp. 143-145). And the fact that a weapon like the drone can destroy the target in a more precise way than another one would be able to do does not mean that it is, at the same time, better able to distinguish who constitutes a legitimate target and who, instead, is not (p. 142). Indeed, Chamayou reflects on the seriousness of the legal dilemma resulting from the use of the drone: the impossibility of attributing responsibility for the act which, divided within an intricate network of people, entities and organizations, is diluted, transforming a war crime into a mere military and industrial accident (p. 211).

Furthermore, while in some war zones - such as Afghanistan - the legality of the war conducted by drones is not questioned, however, it is considerably problematic - both under international and domestic law - in other countries, such as Yemen, Pakistan, and Somalia, with which the United States is not currently at war. Since the warfare drone, warfare conducted by drones, is considered, within the sphere of international law, as a new way of conducting a war, since it eliminates the reciprocity that, in "normal" situations, warfare itself presents, new forms of legislation are needed to regulate its course (Graae & Maurer, 2017, p. 6). On the other hand, it has been argued that the policies implemented by the Obama Administration during its mandate and concerning targeted assassinations have, over the years, increasingly eroded the traditional legal and ethical dissuaders regarding the use of force, creating a deep split between the current American policy on the one hand, and, on the other, the pre-established modality of the use of force according to the law governing armed conflict (Boyle, 2015, p. 122).

Before dealing with the main object of this paragraph, i.e., the complex legal nature of the drone, it is necessary to specify that the two notions belonging to the sphere of international law to which reference will be made are *ius in bello* (law in war or IHL, International Humanitarian Law) and *ius ad bellum* (IHRL, International human rights law). The International Committee of the Red Cross (2010) defines the former as the set of rules governing the manner in which war is conducted; the latter, on the other hand, is the body of law governing the legitimacy of the reasons that led to the outbreak of conflict, whether between two state actors or between a state and a non-state actor. A further logical difference between the two *iuras* lies in the fact that, while *ius in bello* applies only in the case of an actual armed conflict, *ius ad bellum* is applicable both in peace and in war. There are basically three sources of international humanitarian law: the Geneva Conventions of 1949, the Additional Protocols annexed to them, and the Hague Conventions of 1899 and 1907. The *ius ad bellum*, on the other hand, which is implemented through the mechanisms of the Charter of the United Nations, is the result of a progressive development of various international instruments, including the DUDU (the Universal Declaration of Human Rights of 1948), the International Convention on Economic, Social and Cultural Rights and the International Covenant on Civil and Political Rights, both signed in 1966. In particular, Article 2(4) of the Charter of the United Nations (1945, Chapter I, Art. 2, Par. 4) prohibits the threat or use of force against the territorial integrity or political independence of any state. There are two exceptions to this rule (Valitis, n.d., pp. 29-30): authorization by the UN Security Council to intervene to maintain or restore international peace and security; and state self-defense in the event of armed attack. However, the amount of force used in the latter case must be commensurate with the principles of necessity and proportionality that *ius ad bellum* provides for.

In fact, within the legal framework of the IHRL, entry into war involves certain requirements: the declaration of the status of war by a public authority, just cause, the probability of success, the principle of proportionality and, finally, the principle of last resort. Therefore, through the lens of *ius ad bellum*, the lethal use of force is justified - and therefore permitted - only when a subject poses an imminent threat to someone's life (Gaggioli, 2013, p. 17). Therefore, the IHRL prefers a "capture rather than kill" approach to subjects considered as suspects. Some academics (Walch, 2014, pp.63-65) have pointed out that it is precisely because of this that the situation becomes more complicated, since the intention of the drone is rather at the antipodes of the one just mentioned, since it "kills rather than kill", and this totally undermines its compatibility with international human rights law.

This legal aspect acquires the aspect of a real dilemma, if we consider the specific case of remote-piloting aircraft. In fact, it is necessary to underline that, should the single attack operated by the drone be classified as an "armed attack", it would become necessary, for the state that authorized the attack, to justify its action - should the UN Security Council not give its authorization to proceed - as an act of self-defense, since otherwise it would risk being accused of aggression (Casey-Maslen, 2012, pp.601-603), which is defined as such by the UN Security Council (United Nations General Assembly, 1974, Resolution 3314 (XXIX)):

<<Aggression is defined as the use of armed force by a State against the sovereignty, territorial integrity or political independence of another State, or in any other way incompatible with the Charter of the United Nations, as set out in this definition.>>

Everything stated so far is closely related to an important distinction that the IHL makes between IAC and NIAC, respectively International Armed Conflict (International Armed Conflict) between states and Non-International Armed Conflict (Non-International Armed Conflict) conducted between internal factions of the state. In fact, the crucial issue with respect to drone attacks concerns the very identity of the attack conducted by a country that is not - at least officially - at war with the country it decides to attack. In this sense it becomes fundamental to classify the area within which the attack is considered whether it is an armed conflict or not. The International Committee of the Red Cross (2008, p. 1) points out that the international armed conflict is between two or more states, whereas the non-international armed conflict takes place between government forces and non-governmental armed groups, or only between these groups.

It can therefore be inferred that the drone attacks in the territories of Pakistan and Yemen are not classifiable as IACs, since they were not interstate attacks. In fact, taking the first of these two into consideration, it can be concluded (O'Connell, 2010) that there is no legal right to use drones in Pakistan, since not only has Pakistan not attacked the United States - and therefore there is no reason to resort to the law of self-defense - but, furthermore, there are numerous doubts about the respect of the principles that *ius in bello* invokes, such as that concerning the proportionality of the attacks carried out. In any case, even if the instrument of legitimate defense is used, it is fundamental to respect the principles of distinction between fighters and civilians, military necessity, proportionality between the attack carried out and the losses, damage limitation and humanity that the IHL body imposes (Oberson, n.d., p. 7). Therefore, in order to take legal action, drones must at the same time act in respect of the principle of *ius ad bellum* of the use of force as a last resort as well as the fundamental principles of *ius in bello* (Alim, 2018, p. 74). All of the above implies that, since the attack carried out by the drone in another state implies the presence of armed conflict, the drone operator must be considered as a combatant and, therefore, must comply with the norms of international human rights law (p. 76).

In light of all this, if an attack is classified as an armed conflict, then both IHL and IHRL are applicable, and therefore the legality of drone attacks is determined by the applicable *lex specialis*, namely the IHL itself (Gregory, 2013, p. 4). Chamayou (2015, p. 171) states that, therefore, an attack conducted by drones can only be defined as legal in two ways: either the attacks are considered as part of IHRL, and therefore drones are obliged by law to proportionate their use of force (which is impossible for a drone to do), or drones fall within the scope of IHL, even if the laws of the latter do not apply when there is no conflict situation, as in fact it is in the territories of Pakistan and Yemen.

Moreover, the "persistent presence" of the drone, which allows it to intervene at any time within the airspace of another state, has been placed on the same level as an air occupation. In fact, the drone can both respect the principle of territorial sovereignty - since the action takes place along a vertical space and no soldiers are deployed in the field, horizontally territorial sovereignty is respected - and transgress it, if there is no legal justification for the incursion of a drone into the territory of another state (Munro, 2014, pp. 238-239).

One of the arguments which concerns drones and which, at a legal level, arouses greater controversy is, undoubtedly, the practice of the double-tap strike - literally, the "double strike" - by the CIA, a practice in which the operators of the drones strike those who are preparing to rescue the targets of a first attack operated by the drone, or, in any case, subjects who organize the funeral. The justification for the double hits lies in the assumption that the persons providing aid to the affected person are either familiar to the cause of the latter or to the terrorists themselves (Gusterson, 2017): in short, in both cases they would be dangerous subjects. In this regard, some scholars (Cavallaro et al., 2012, p. 76) have stated that:

<< [...] Not only does this practice call into question the extent to which secondary attacks comply with the fundamental rules of international humanitarian law regarding distinction, proportionality and precautions, but it also potentially violates specific legal safeguards for medical and humanitarian personnel and the injured. As experience in international law has shown, intentional attacks on first responders can constitute war crimes.>>

And it is precisely as "war crimes" that some studies (Orozobekova, 2015, p. 26) categorize *signature strikes*, namely the targeted murders that were dealt with in the second chapter, and *double-tap strikes* since, as Amnesty International states (2013, p. 30), "deliberately attacking civilians who are providing aid to the wounded or the wounded themselves constitutes a war crime".

Therefore, one of the major criticisms of the practice of targeted killing (Benson, 2014, pp. 36-37) concerns the principles of proportionality and distinction in international humanitarian law. In fact, the principle of proportionality implies that, even if the existence of a possible side effect is admitted, the latter must be proportionate, that is adequate and not excessive, in the light of the military advantage anticipated by the attack itself, and since the drone does not possess the capacity to consider the cultural context, this could lead to a side effect - in this case the killing of civilians - that is not proportionate. Moreover, the fact that there are non-civilians in a civilian area does not automatically imply that the area itself becomes a military target, let alone that the civilians present in that area lose their civilian status (p. 38), as is further underlined by the First Additional Protocol to the Geneva Conventions (1949, Art. 50, Par. 3). Furthermore, the International Committee of the Red Cross (2010) has pointed out that the psychological traumas that civilian killings entail constitute a lack of respect by political actors for their obligations under international humanitarian law.

In its defense, the U.S. administration promoted the legal drone-based attacks on the grounds of self-defense against al-Qaeda, which attacked the United States on 11 September 2001. Furthermore, the United States claims that Article 2(4) of the UN Statute only prohibits the unilateral use of force and, therefore, if the state in which the attack took place gives its consent, the attack itself no longer violates the sovereignty of the state in question (Purkiss & Serle, 2017).

The other motivation promoted by the Americans is that a state can use force when the aim is to prevent an attack by a non-state actor (Ali & Khan, 2017, pp. 190 and 195). However, it has been repeatedly pointed out that some states, such as Pakistan, actually regard US attacks in the FATA region as a violation of their territorial integrity and political sovereignty, and for two very specific reasons. First, Pakistan claims that it has not undertaken any kind of attack against the United States, either directly or indirectly; second, it has not even given its consent in Washington to attack Pakistan through the use of drones. Therefore, this logic would imply that it is impossible for the US to use the motivation of self-defense to justify its drone attacks in this state (p. 202). In this regard, supporters of military drones (Vogel, 2011, p. 128) claim that there is no evidence that remotely piloted aircraft cause more damage than traditional bombardments. In addition, the State Department's legal counsel between 2009 and 2013, Harold Hongju Koh (2010), highlighted the fact that US practices involving the use of remotely piloted aircraft - including lethal operations - comply with all applicable laws, including the laws of war. On the other hand, the President of the Red Cross Committee, Peter Maurer (2013), underlined that, according to the canons of international humanitarian law, drones are not explicitly prohibited, nor are they considered as intrinsically indiscriminate or treacherous instruments; from this point of view, they are, in practice, comparable to any weapon that is launched from a manned aircraft.

In conclusion, some academics (Radsan & Murphy, 2011, p. 1205) have stressed the fact that, if measures are not taken to prevent the use of the instrument of targeted assassinations against suspected terrorists (especially by the United States of America), it is likely that the fate of this practice will continue. Furthermore, there is also the possibility that a consensus can be reached on the legality of targeted assassinations and their compatibility with the principles of international humanitarian law, as pointed out by a well-known international law expert, Nils Melzer (2008, p. 9):

<<Today, the practice of targeted homicide is on the verge of escaping the dark realm of semi-legality and non-responsibility, and gradually acquiring legitimacy as a method of counter-terrorism and "surgical" warfare.>>

On the other hand, all the questions that have been taken into consideration in this paragraph contribute to create what Crosston (2014, p. 7) defines as an "incredibly slippery ethical slope", where states have different evaluation parameters to decide whether a target is legal or not and which military attack is the result of an imminent military necessity and which is not. An act of killing that is authorized and carried out at a distance has been interpreted as murder rather than honorable combat; moreover, since this has often been considered 'cowardly', it could imply the unfortunate result of producing even more support for terrorist organizations, rather than undermining their credibility; finally, since investment in the production and trade of remotely piloted aircraft is on the rise, the use of the latter poses numerous challenges in terms of respect for and implementation of international humanitarian law (McDonnell, 2012, pp. 309-316).

Ultimately, it will become essential to solve the dilemma of the "drone myth" (Brunstetter, 2011, p. 3) and the consequences that this may have for those who will use drones in the future are particularly difficult to predict, especially because drones, like all the weapons and technologies that preceded them, are anything but infallible and, in their use, a multiplicity of factors must be taken into account, among which stand out the limitations deriving from the human factor (p. 8), that is the presence of the operator - who, even if not physically in the field, controls the drone - which will be the subject of the fourth and final paragraph.

4. THE HUMAN FACTOR WITHIN THE USE OF DRONES

The problem of the Human Factor with respect to remotely piloted aircraft is intrinsically linked to the figure of the operator driving the drone. And this is easily deduced from the fact that it is human intelligence that decides to hit a target, and not the machine (Valitis, n.d., p. 12). In fact, the fact that the drone operator and the drone itself are in two different places at the same time implies numerous complications, including the loss of sensory cues useful for the direction of flight, delays in control and communication cycles and difficulties in scanning the visual environment surrounding the vehicle itself (McCarley & Wickens, 2005, p. 2). From the operator's point of view, the bodies are reduced to small dark figures, while the houses form the background, and an exploding bomb represents a cloud of smoke that is fading (Tognocchi, 2018, p. 28).

Therefore, the problem that demands more attention - if we consider the relationship between a drone and its operator - is that of the long distance between these two elements, or dehumanization (Pešek, 2016, p. 39). Especially worthy of attention is the fact that this relationship helps to paint the existing reality from the point of view of the drone itself as similar to that of a video game, where there is no physical contact with the enemy and, therefore, there is no real risk of being injured by the latter (p. 40).

And the consequences of this for the drone operator are manifold, especially on a psychophysical level. Some military personnel who have served as drone operators in the past (Hussain, 2015) have reported numerous episodes of violence and derision that they have witnessed, especially among their colleagues. Furthermore, numerous academics (Gregory, 2014) have analyzed the problems that the operator faces in the course of his daily tasks, paying particular attention to the fact that he constantly has to make a "switch" between different worlds: on the one hand the working world, where he is an operator who carries out orders; on the other hand, the personal and private world, where he cultivates passions, interests, and most probably a family.

What was initially referred to (Franz & Ladewig, 2017, p. 35) as "applied experimental psychology" and, since the 1940s, as "human factors studies", encompasses a whole series of examinations based on occupational aptitude tests for pilots. Nowadays, human factors problems (some practical examples will be given shortly), also referred to as sensation and perception problems, have proven to be crucial for engineers dealing with the human-machine interface (Treccani encyclopaedia), especially since the aim is, on a technical level, to optimize the use of human capabilities (Byman, 2013, pp. 35-36), so as to minimize the susceptibility of technical systems to disturbance by the human factor itself (Di Loreto, 2006, p. 65).

As far as remotely piloted aircraft are concerned, automation, i.e. the ability of a machine to perform a function that was previously carried out by humans, is of primary importance: in the case of UAVs, it is the functions relating to direct control of the aircraft and the performance of operations such as target recognition that are automated (Cooke et al., 2006).

An equally important role is played by the operator of the remotely piloted aircraft in the control station. The operator is not only responsible for guiding the drone, but also for providing the drone with the commands to be executed and analyzing the information collected during the various missions. As far as the optimal functioning of remotely piloted aircraft is concerned, a great deal of research has been carried out with the aim of correcting these problems. The research that I have specifically analyzed in this context is that of the engineer Giovanni Miranda, who proposes in his research a human-machine interface simulator for the control of Unmanned Aerial Vehicles, precisely with the aim of intervening on the human factor, so as to optimize the use of the aircraft. In fact, Dr. Miranda points out (2008, p. 29) that, during the design of UAVs, research focused more on the technological aspect to be developed, rather than on the human factor (for instance level of automation, sensory isolation and information saturation), relegating the latter to the margins of the research itself. This has resulted in human error causing numerous accidents (pp. 27-28), both as a primary and secondary cause (i.e., Miranda explains, when the operator is unable to remedy automation failures). In this respect, the percentage of accidents attributable to Human Factors, which varies depending on the drone in question, ranges from 21% to 68% (Williams, 2004, p. 12).

In the course of his doctoral research, Dr Miranda (2008, p. 28) presented a prototype simulation system through which it is possible to simulate different types of UAV system missions, with the aim of stimulating equally different interactive modes between man and machine. The results of this simulator proved to be very interesting (pp. 120-121). In fact, the characteristic presence of the two displays was able to provide the operator with easy access, both direct and indirect, to all the information he needed to understand and be aware of the UAV's motion and changes in the scenario, so that he was able - without making an excessive mental effort - to make reliable predictions about the status of the aircraft and the scenario in the short term. Also positive was the representation of all mission-relevant information on a single screen, the 3D Virtual Display, which allowed the operator to keep the primary information under constant control and, at the same time, to call up other information through the implementation of simple operations.

The test results were equally encouraging. In fact, not only does the simulator promoted by Dr. Miranda within his research encompass all the fundamental criteria that make it "usable" at an ergonomic level, namely effectiveness, efficiency, learnability and memorability (Somma, 2020), but it also does not encourage the commission of errors and, even if errors are committed, the user is able to make them null and void (Miranda, 2008, p. 123).

CONCLUSION

Remotely piloted aircraft - or UAVs - are a growing sector and are destined to continue to play a major role on the international relations scene. It is from this premise that my research has moved. In fact, this investigation started from the very beginning of the history of UAVs, outlining a path that has led them to occupy the important role they hold today. There are many questions surrounding UAVs, many of which - in my opinion - are destined to remain unanswered. At the end of this research, what I think is self-evident is that drones are a double-edged sword that must be used with the utmost caution. I am not questioning the importance that this technology has and will have, especially on the international stage (but not only); however, I think that the use that has been made of it should not, in certain contexts, be repeated in an imprudent manner. From this point of view, research in the legal field, which has particularly fascinated me, teaching me numerous specific notions on this subject, speaks clearly: the debate on the correct use of drones that is compatible with international law is far from being resolved.

This same research leaves an open end, so to speak. It has been shown that the usefulness and importance of remotely piloted aircraft are not questioned, at least not to date; it is the questions about their daily use, as well as the motivations behind their use, and the future consequences of their use - both ethically and legally - that are difficult to answer adequately.

No matter how far-fetched the predictions may be, only time will provide a reliable answer to these questions. In the meantime, information and the continuous study of remotely piloted aircraft are the weapons each of us possesses to better understand the advantages and dangers of this incomparable technology.

ACKNOWLEDGMENT

I would like to express my deepest appreciation to my thesis advisor, Professor Luisa Dall'Acqua: not only did she play a decisive in guiding me through the construction of the entire master's dissertation, but she was also supportive of my work from the very beginning.

I am also deeply indebted to Giovanni Miranda, my thesis co-rapporteur, who taught me all I needed to know about UAVs and automation: this work would not exist, if it were not for you.

REFERENCES

Additional Protocol to the Geneva Conventions of 12 August 1949 for the Protection of Victims of International Armed Conflicts, adopted on 8 June 1977. (n.d.). https://www.admin.ch/opc/it/classified-compilation/19770112/index.html

Agency France-Presse. (2019). US strikes in Somalia won't stop Shabaab, says US general. *Daily Nation*. https://www.nation.co.ke/news/world/US-General--Strikes-in-Somalia-won-t-stop-Al-Shabab/1068-4972234-h7vygv/index.html

Ali, S. & Khan, A. A. (2017). Drone strikes, Logic of Self-defense and Violation of State Sovereignty. *Pakistan Journal of History and Culture, 38*(2), 187-206.

Alim, A. (2018). International Armed Conflict and Drone Warfare: Quest for a New Apace in international Humanitarian Law. *International Journal of Law, Humanities & Social Science, 2*(4), 70-82.

Alley, R. (2013). *The Drone Debate, Sudden Bullet or Slow Boomerang?* Discussion Paper in Centre for Strategic Studies: New Zealand, Victoria University of Wellington.

Amazon.com. (n.d.). https://www.amazon.com/Amazon-Prime-Air/b?ie=UTF8&node=8037720011

American Congress. (2001). *Public Law 107–40—SEPT. 18, 2001, 107th Congress, Authorization for Use of Military Force. 50 USC 1541 note.*

Amnesty International. (2013). *"Will I be next?" US Drone strikes in Pakistan.* Amnesty International Publications.

BensonK. (2014). "Kill 'em and Sort it Out Later": Signature Drone Strikes and International Humanitarian Law. *Pacific McGeorge Global Business & Development Law Journal, 27*(1). https://ssrn.com/abstract=2465620

Bergen, P., & Rowland, J. (2012). *Obama ramps up covert war in Yemen.* CNN. https://edition.cnn.com/2012/06/11/opinion/bergen-yemen-drone-war/index.html

Bergen, P., Sterman, D., Salyk-Virk, M., Sims, A., & Ford, A. (n.d.a). *The War in Yemen.* New America. https://www.newamerica.org/in-depth/americas-counterterrorism-wars/us-targeted-killing-program-yemen/

Bergen, P., Sterman, D., Salyk-Virk, M., Sims, A., & Ford, A. (n.d.b). *The War in Somalia.* New America. https://www.newamerica.org/international-security/reports/americas-counterterrorism-wars/the-war-in-somalia/

Bergen, P., Sterman, D., Salyk-Virk, M., Sims, A., & Ford, A. (n.d.c). *The Drone War in Pakistan.* New America. https://www.newamerica.org/in-depth/americas-counterterrorism-wars/pakistan/

Bowden, M. (2013). The Killing Machines, How to think about drones. *The Atlantic.* https://www.theatlantic.com/magazine/archive/2013/09/the-killing-machines-how-to-think-about-drones/309434/

Boyle, M. J. (2015). The legal and ethical implications of drone warfare. *International Journal of Human Rights, 19*(2), 105–126. doi:10.1080/13642987.2014.991210

Brooks, D. (2013). Florence and the Drones. *The New York Times.* https://www.nytimes.com/2013/02/08/opinion/brooks-florence-and-the-drones.html

Brooks, R., & Hastings, M. (2012). The Rise of the Killer Drones: How America Goes to War in Secret. *Rolling Stone*. https://www.rollingstone.com/politics/politics-news/the-rise-of-the-killer-drones-how-america-goes-to-war-in-secret-231297/

Brooks, R. (2013). *10 ways to fix the drone war*. Foreign Policy. https://foreignpolicy.com/2013/04/11/10-ways-to-fix-the-drone-war/

Brunstetter, D., & Braun, M. (2011). The implications of drones on the just war tradition. *Ethics & International Affairs, 25*(3). https://www.cambridge.org/ core/ journals/ ethics- and- international- affairs/ article/ abs/ implications- of- drones-on- the- just- war- tradition/ 97ABF 476B84 94CC44 A71E011 DD8B7600

Bush, G. W. (2001). *Address Before a Joint Session of the Congress on the United States Response to the Terrorist Attacks of September 11*. The American Presidency Project. https://www.presidency.ucsb.edu/ documents/ address- before- joint- session-the- congress- the- united- states- response- the- terrorist- attacks

Business Insider Intelligence. (2020). *Drone market outlook: industry growth trends, market stats and forecast*. Business Insider. https://www.businessinsider.com/drone-industry-analysis-market-trends-growth-forecasts?IR=T

Byman, D. L. (2013). *Why Drones Work: The Case for Washington's Weapon of Choice*. Brookings. https://www.brookings.edu/articles/why-drones-work-the-case-for-washingtons-weapon-of-choice/

Byrne, E. F. (2015). Making Drones to Kill Civilians: Is it Ethical? *Journal of Business Ethics*, 1-13. . doi:10.100710551-015-2950-4

Cambridge Dictionary. (n.d.). https://dictionary.cambridge.org/it/dizionario/inglese/uav

Casey-Maslen, S. (2012). Pandora's box? Drone strikes under jus ad bellum, jus in bello, and international human rights law. *International Review of the Red Cross, 94*(886), 597–625. doi:10.1017/S1816383113000118

Cavallaro, J., Sonnenberg, S., & Knuckey, S. (2012). Living Under Drones: Death, Injury and Trauma to Civilians from US Drone Practices in Pakistan. Stanford: International Human Rights and Conflict Resolution Clinic, Stanford Law School; New York: NYU School of Law, Global Justice Clinic.

Cavoukian, A., & Office of the Information and Privacy Commissioner/Ontario. (2012). *Privacy and drones: Unmanned aerial vehicles*. Ontario, Canada: Information and Privacy Commissioner of Ontario, Canada.

Chamayou, G. (2015). *A Theory of the Drone*. The New Press.

Clark, K. (2017). *Drone warfare 1: Afghanistan, birthplace of the armed drone*. Afghanistan analysts Network. https://www.afghanistan-analysts.org/ en/reports/ international-engagement/ drone-warfare- 1-afghanistan- birthplace-of- the-armed-drone/

Cloud, D. S. (2010). CIA drones have broader list of targets. *Los Angeles Times*. https://www.latimes.com/world/la-xpm-2010-may-05-la-fg-drone-targets-20100506-story.html

Cooke, N. J., Pringle, H., Pedersen, H., & Connor, O. (2006). *Human Factors of Remotely Piloted Vehicles. Advances in Human Performance and Cognitive Engineering Research Series*. Elsevier.

Cronin, A. K. (2013). *Why Drones Fail: When Tactics Drive Strategy*. Foreign Affairs. https://www.foreignaffairs.com/articles/somalia/2013-06-11/why-drones-fail

Crosston, M. (2014). Pandora's Presumption: Drones and the Problematic Etchis of Techno-War. *Journal of Strategic Security*, 7(4), 1–24. doi:10.5038/1944-0472.7.4.2

Currier, C., & Maass, P. (2015). *Firing blind: Flawed intelligence and the limits of drone technology: The drone papers*. The intercept. https://theintercept.com/drone-papers/firing-blind/

Di Loreto, M. (2006). *Impiego Joint e Combined di Unmanned Aerial Vehicles (UAV): Stato dell'arte e prospettive future di impiego*. eMiSS, Centro Militare di Studi Strategici.

Feroz, E. (2020). *Death by drone: America's vicious legacy in Afghanistan*. Foreign Policy. https://foreignpolicy.com/2020/03/27/afghanistan-drones-america-legacy-taliban/

Fischer, D. (2014). *Scare or care: the War on Terror 2001-2013*. Ruprecht-Karls-Universität Heidelberg, Heidelberg Center for American Studies.

Fischer, D. (2014). Scare or care: the War on Terror 2001-2013. Ruprecht-Karls-Universität Heidelberg, Heidelberg Center for American Studies.

Franz, N., & Ladewig, R. (2017). Synthetic Reality and Blind Fight: The Operationalization of Perception. *Politik*, *20*(1), 30–44.

Gaggioli, G. (2013). *The use of force in armed conflicts, Interplay between the conduct of hostilities and law enforcement of paradigms*. International Committee of the Red Cross' Expert Meeting.

Gettinger, D. (2020). The Soleimani killing and 5 things to know about drones in Iraq. *Bulletin of the Atomic Scientists*. https://thebulletin.org/2020/01/soleimani-and-beyond-5-ways-that-drones-have-destabilized-iraq/#

Graae, A. I., & Maurer, K. (2017, March). Introduction: Debating Drones: Politics, Media, and Aesthetics. *Politik, 20*(1). Advance online publication. doi:10.7146/politik.v20i1.27640

Greenwald, G. (2012). Obama moves to make the War on Terror permanent. *The Guardian*. https://www.theguardian.com/commentisfree/2012/oct/24/obama-terrorism-kill-list

Gregory, D. (2014). *Drone geographies*. Radical Philosophy. https://www.radicalphilosophy.com/article/drone-geographies

Gregory, T. A. (2013). *Drones: Mapping the Legal Debate*. Università di Auckland.

Gusterson, H. (2017). *Drone warfare*. American Institute of Physics. https://aip.scitation.org/doi/abs/10.1063/1.5009234 doi:10.1063/1.5009234

Hambling, D. (2019). *Autonomous killer drones set to be used by Turkey against Syria*. New Scientist. https://www.newscientist.com/article/2217171-autonomous-killer-drones-set-to-be-used-by-turkey-in-syria/

Hansen, S. J. (2019). *Somalia drone strikes are a potent weapon, but not the game changer*. The Conversation. https://theconversation.com/somalia-drone-strikes-are-a-potent-weapon-but-not-the-game-changer-111845

Hassanalian, M., & Abdelkefi, A. (2017). Classifications, applications, and design challenges of drones: A review. *Progress in Aerospace Sciences, 91*, 99–131. doi:10.1016/j.paerosci.2017.04.003

Hastings, M. (2012). *The Rise of the Killer Drones: How America Goes to War in Secret*. Rolling Stone. https://www.rollingstone.com/politics/politics-news/the-rise-of-the-killer-drones-how-america-goes-to-war-in-secret-231297/

History.com Editors. (2019). *A Timeline of the U.S.-Led War on Terror*. A&E Television Networks. https://www.history.com/topics/21st-century/war-on-terror-timeline

Hoffman, F., & Kalikow, E. (2013). *To drone or not to drone*. Foreign Policy Research Institute. https://www.fpri.org/article/2013/08/to-drone-or-not-to-drone/

Hussain, M. (2015). *Former drone operators say they were "horrified" by cruelty of assassination program*. The Intercept. https://theintercept.com/2015/11/19/former-drone-operators-say-they-were-horrified-by-cruelty-of-assassination-program/

International Committee of the Red Cross. (2008). *How is the Term "Armed Conflict" Defined in International Humanitarian Law?* https://www.icrc.org/en/doc/resources/documents/article/other/armed-conflict-article-170308.htm#:~:text=International%20humanitarian%20law%20distinguishes%20two,or%20between%20such%20groups%20only.

International Committee of the Red Cross. (2010a). *IHL and other legal regimes*. ICRC. https://www.icrc.org/en/document/ihl-and-other-legal-regimes

International Committee of the Red Cross. (2010b). *Protection of the civilian population*. https://www.icrc.org/en/doc/what-we-do/protecting-civilians/overview-protection-civilian-population.htm

Keene, S. D. (2015). *Lethal and legal? The ethics of drone strikes*. Strategic Studies Institute and U.S. Army War College Press.

Koh, H. H. (2010). *Annual Meeting of the American Society of International Law*. https://2009-2017.state.gov/s/l/releases/remarks/139119.htm

Lee, D. (2019). *Amazon to deliver by drone 'within months'*. BBC. https://www.bbc.com/news/technology-48536319

Lee-Morrison, L. (2014). *State of the Art Report on Drone-based Warfare*. Division of Art History and Visual Studies, Department of Arts and Cultural Sciences, Lund University.

Lewis, M. W. (2012). Drones and the boundaries of the battlefield. *Texas International Law Journal*, *47*(2), 293–314.

Lister, T., & Cruickshank, P. (2011). *Anwar al-Awlaki: al Qaeda's rock star no more*. CNN. https://edition.cnn.com/2011/09/30/world/meast/analysis-anwar-al-awlaki/index.html

Markets And Markets. (2018). *Military Drones Market by Type, Application (ISRT, Delivery and Transportation, Combat Operations, Battle Damage Management), Range (VLOS, EVLOS, BLOS), Propulsion Type, Launching Mode, Endurance, MTOW, Region - Global Forecast to 2025*. Markets and Markets. https://www.marketsandmarkets.com/Market-Reports/military-drone-market-221577711.html

Masters, J. (2011). *Closing Guantanamo?* Council on Foreign Relations. https://www.cfr.org/backgrounder/closing-guantanamo

Matějka, S. (2014). *Unmanned Aerial Vehicles in US National Security Policy, New Face of War on Terror.* Univerzita Karlova V Praze, Fakulta Sociálních Věd, Institut Mezinárodních Studií, Katedra Amerických Studií, Praga.

Maurer, P. (2013). *The use of armed drones must comply with laws.* International Committee of the Red Cross. https://www.icrc.org/en/doc/resources/documents/interview/2013/05-10-drone-weapons-ihl.htm

Mazzetti, M., & Sanger, D. E. (2009). *Obama expands missile strikes inside Pakistan.* The New York Times. https://www.nytimes.com/2009/02/21/washington/21policy.html

McCarley, J. S., & Wickens, C. D. (2005). *Human Factors Implications of UAVs in the National Airspace.* Institute of Aviation, Aviation Human Factors Division, University of Illinois at Urbana-Champaign.

McDonnell, T. M. (2012). Sow What You Reap? Using Predator and Reaper Drones to Carry Out Assassinations or Targeted Killings of Suspected Islamic Terrorists. *The George Washington International Law Review, 44*(243), 243-316. https://digitalcommons.pace.edu/lawfaculty/859/

McKenna, A. (2016). The public acceptance challenge and its implications for the developing civil drone industry. In The future of drone use: opportunities and threats from ethical and legal perspectives. T.M.C. Asser Press. doi:10.1007/978-94-6265-132-6_18

McManus, D. (2009). *The cost of killing by remote control.* Los Angeles Times. https://www.latimes.com/archives/la-xpm-2009-may-03-oe-mcmanus3-story.html

Melzer, N. (2008). *Targeted Killings in International Law.* Oxford University Press Inc.

Miller, G. (2011). Under Obama, an emerging global apparatus for drone killing. *The Washington Post.* https://www.washingtonpost.com/ national/national-security/ under-obama- an-emerging- global-apparatus-for-drone-killing/ 2011/12/13/ gIQANPdILP_story.html

Miller, G., DeYoung, K., Whitlock, C., & Tate, J. (2012). *Plan for hunting terrorists signals U.S. intends to keep adding names to kill lists.* The Washington Post. https://www.washingtonpost.com/ world/ national-security/ plan-for- hunting-terrorists-signals-us- intends-to- keep-adding- names-to- kill-lists/ 2012/10/23/ 4789b2ae-18b3- 11e2- a55c- 39408f be6a4b_ story.html

Mir, A. (2018). *The U.S. drone war in Pakistan revisited.* Stanford, Freeman Spogli Institute for International Studies. https://fsi.stanford.edu/news/us-drone-war-pakistan-revisited

Miranda, G. (2008). *Simulatore di interfaccia uomo-macchina per il controllo di UAV.* PhD, Alma Mater Studiorum, Università di Bologna.

Munro, C. A. O. (2014). Mapping the vertical battlespace: towards a legal cartography of aerial sovereignty. *London Review of International Law, 2*(2), 233-261. . doi:10.1093/lril/lru008

Nakashima, E., & Whitlock, C. (2011). *With Air Force's Gorgon Drone 'we can see everything'.* The Washington Post. https://www.washingtonpost.com/wp-dyn/content/article/2011/01/01/AR2011010102690.html?sid=ST2011012204147

Netti, E. (2020). *Droni, il mercato prende quota.* Il Sole 24 ORE. https://www.ilsole24ore.com/art/droni-mercato-prende-quota-ACQSyRIB

O'Connell, M. E. (2010). The International Law of Drones. *American Society of International Law, 14*(37). https://www.asil.org/insights/volume/14/issue/37/international-law-drones

Oberson, B. (n.d.). *Diritto Internazionale Umanitario.* Comitato Internazionale della Croce Rossa. https://www.google.com/ search?q= Oberson% 2C+B.+ Diritto+ Internazionale+ Umanitario% 2C+in+% 3C%3 C+Comitato+ Internazionale+ della+ Croce+ Rossa% 3E% 3E% 2C+p.+ 7.&rlz= 1C5CHFA _enAT892 AT892& sxsrf= ALeK k00Wny Sgzlup GdQ9S8 hBD7tj GJoDkQ %3A161 7092965 509&ei= ZeFiYM CqHsW TkwXZu IbgAg&oq= Oberson% 2C+B.+ Diritto+ Internazionale+ Umanitario%2 C+in+% 3C% 3C+Comitato+ Internazionale+ della+ Croce+ Rossa% 3E% 3E% 2C+p.+7. &gs_lcp= Cgdnd 3Mtd2 l6EA MyBA gAEEc yBAgA EEcyB AgAEE cyBAg AEEcy BAgAE EcyBA gAEEc yBAgA EEcyBA gAEEdQ 3FFYp11 g415oAH ADeA CAAQC IAQCS AQCYAQ GgAQ GqAQdn d3Mtd2l6y AEIwAEB &sclient= gws-wiz& ved= 0ahUKE wiA8Z vezNfv AhXFya QKHVm cASwQ 4dUDC A0&uact=5

Office of the Secretary of Defense of the United States of America (2005). *Unmanned Aircraft Systems Roadmap 2005-2030.* Author.

Orban, V. (2014). I droni nelle operazioni di intelligence e anti-terrorismo in Yemen e Pakistan. *Osservatorio di Politica Internazionale (BloGlobal – Lo sguardo sul mondo), 2,* 1-12.

Orban, V. (2015). *Il mercato globale dei droni: profili militari e prospettive geopolitiche*, in *Intelligenza artificiale nei nostri cieli – Neuron, il drone europeo*. NoGeoingegneria. https://www.nogeoingegneria.com/tecnologie/spazio/intelligenza-artificiale-nei-nostri-cieli-neuron-il-drone-europeo/

Orozobekova, C. (2015). The US Drone Policy Under International Law. *Spotlight on Regional Affairs, 34*(11-12). http://www.irs.org.pk/spotlight/spnd2015.pdf

Pešek, P. (2016). *The Role of Drones in Contemporary Warfare*. Metropolitan University Prague, International Relations and European Studies.

Priest, D., & Arkin, W. (2011). *Top Secret America: The Rise of the New American Security State*. Little, Brown and Company.

Purkiss, J., & Serle, J. (2017). *Obama's covert drone war in numbers: ten times more strikes than Bush*. The Bureau of Investigative Journalism. https://www.thebureauinvestigates.com/stories/2017-01-17/obamas-covert-drone-war-in-numbers-ten-times-more-strikes-than-bush

Radsan, A. J., & Murphy, R. (2011). Measure Twice, Shoot Once: Higher Care for CIA-Targeted Killing. *University of Illinois Law Review*, 1201–1242.

Read, R. (2020). *World's most feared drone: CIA's MQ-9 Reaper killed Soleimani*. Washington Examiner. https://www.washingtonexaminer.com/policy/defense-national-security/worlds-most-feared-drone-cias-mq-9-reaper-killed-soleimani

Sabbagh, D. (2019). *Killer drones: how many are there and who do they target?* The Guardian. https://www.theguardian.com/news/2019/nov/18/killer-drones-how-many-uav-predator-reaper

Sauer, F., & Schörnig, N. (2012). Killer drones: The 'silver bullet' of democratic warfare? *Sage Journals, Security Dialogue, 43*(4), 363-380. https://journals.sagepub.com/doi/full/10.1177/0967010612450207

Serle, J., & Sargand, P. (2015). *Afghanistan: Reported US air and drone strikes 2015*. The Bureau of Investigative Journalism. https://www.thebureauinvestigates.com/drone-war/data/get-the-data-a-list-of-us-air-and-drone-strikes-afghanistan-2015

Shachtman, N. (2009). *CIA Chef: Drones 'Only Game in Town' for Stopping Al Qaeda*. Wired. https://www.wired.com/2009/05/cia-chief-drones-only-game-in-town-for-stopping-al-qaeda/

Shaw, I., & Akhter, M. (2014, April 4). The dronification of state violence. *Critical Asian Studies, 46*(2), 211–234. doi:10.1080/14672715.2014.898452

Shaw, I. G. R. (2014), *The Rise of the Predator Empire: Tracing the History of U.S. Drones*. Understanding Empire. https://understandingempire.wordpress.com/2-0-a-brief-history-of-u-s-drones/

Shewan, D. (2020). *How to Do a SWOT Analysis for Your Small Business (with Examples)*. The WordStream Blog. https://www.wordstream.com/blog/ws/2017/12/20/swot-analysis

Somma, I. (2020). *Esperienza d'uso vs usabilità*. Telemaco. https://www.telemaco.it/blog/esperienza-di-uso-vs-usabilita/

Stewart, P. (2016). *Exclusive: Obama approves broader role for U.S. forces in Afghanistan*. Reauters. https://uk.reuters.com/article/us-usa-afghanistan-obama-exclusive-idUKKCN0YW055

The Free Dictionary. (n.d.). https://www.thefreedictionary.com/unmanned+aerial+vehicle

The White House. (2009). *Promise kept*. The White House, President Barack Obama. https://obamawhitehouse.archives.gov/iraq

Tognocchi, M. (2018). *Drone warfare and the metamorphosis of battlefield: security, space and technology*. Leiden University.

Treccani online encyclopaedia. (n.d.a). *Interfaccia*. https://www.treccani.it/enciclopedia/interfaccia/

Treccani online encyclopaedia. (n.d.b). *SWOT, analisi*. https://www.treccani.it/enciclopedia/analisi-swot_%28Dizionario-di-Economia-e-Finanza%29/

United Nations. (1945). *Statute of the United Nations*. https://www.un.org/en/sections/un-charter/chapter-i/index.html

United Nations General Assembly. (1974). *Resolution 3314 (XXIX), Definition of Aggression*. University of Minnesota. Human Rights Library.http://hrlibrary.umn.edu/instree/GAres3314.html#:~:text=Aggression%20is%20the%20use%20of,set%20out%20in%20this%20Definition

Vainauskas, A. (2014). *The Strategic Costs and Consequences of the U.S. Drone Warfare*. https://www.academia.edu/7532477/The_Strategic_Costs_and_Consequences_of_the_U.S._Drone_Warfare_DRAFT_

Valitis, D. (n.d.). *The Current & Future Use of Armed Drones: Analysing the Controversy*. https://www.academia.edu/11476850/The_Current_and_Future_Use_of_Armed_Drones_Analysing_the_Controversy

VogelR. J. (2011). Drone Warfare and the Law of Armed Conflict. *Denver Journal of International Law and Policy, 39*(1), 101-138. https://ssrn.com/abstract=1759562

von Clausewitz, C. (1832). *On War* (M. Howard & P. Paret, Trans.). Oxford University Press.

Walch, S. (2014). *US Drone operations against Al Qaeda*. Oxford Brookes University.

Warrell, H. (2020). *From Desert Storm to Soleimani: how US drone warfare has evolved*. Financial Times. https://www.ft.com/content/6346dd78-322d-11ea-9703-eea0cae3f0de

Weiner, T. (2005). *New Model Army Soldier Rolls Closer to Battle*. The New York Times. https://www.nytimes.com/2005/02/16/technology/new-model-army-soldierrolls-closer-to-battle.html?auth=login-facebook&searchResultPosition=1

Williams, K. W. (2004). *A Summary of Unmanned Aircraft Accident/Incident Data: Human Factors Implications*. Civil Aerospace Medical Institute, Federal Aviation Administration, Oklahoma City, OK.

Wilson Center. (2012). *The Efficacy and Ethics of U.S. Counterterrorism Strategy – Transcript of Remarks by John O. Brennan*. Wilson Center. https://www.wilsoncenter.org/event/the-efficacy-and-ethics-us-counterterrorism-strategy

Woods, C. (2011). *US claims of "no civilian deaths" are untrue*. The Bureau of Investigative Journalism. https://www.thebureauinvestigates.com/stories/2011-07-18/us-claims-of-no-civilian-deaths-are-untrue

Woods, C., & Ross, A. (2012). *Revealed: US and Britain launched 1,200 drone strikes in recent wars*. The Bureau of Investigative Journalism. https://www.thebureauinvestigates.com/stories/2012-12-04/revealed-us-and-britain-launched-1-200-drone-strikes-in-recent-wars

Zambernardi, L. (2010, July). Counterinsurgency's Impossible Trilemma. *The Washington Quarterly, 33*(3), 21–34. doi:10.1080/0163660X.2010.492722

Zeleny, J. (2007). *As Candidate, Obama Carves Antiwar Stance*. The New York Times. https://www.nytimes.com/2007/02/26/us/politics/26obama.html

Zenko, M. (2011). *Who can't America kill?* Council on Foreign Relations. https://www.cfr.org/blog/who-cant-america-kill

Zenko, M. (2012a). *How many civilians are killed by U.S. drones?* Council on Foreign Relations. https://www.cfr.org/blog/how-many-civilians-are-killed-us-drones

Zenko, M. (2012b). *Targeted Killings and Signature Strikes*. Council on Foreign Relations. https://www.cfr.org/blog/targeted-killings-and-signature-strikes

Zenko, M. (2012c). *Daniel Klaidman's Revelations*. Council on Foreign Relations. https://www.cfr.org/blog/daniel-klaidmans-revelations

KEY TERMS AND DEFINITIONS

Predator (Drone): MQ-1 Predator is a long-endurance, medium-altitude unmanned aircraft system for surveillance and reconnaissance missions. Presented within a demonstration within the United States Air Force, after the 9/11 attacks Predator was equipped with Hellfire missiles and used during the missions in Afghanistan.

SWOT (Analysis): The acronym indicates the *Strengths*, *Weaknesses*, *Opportunities*. and *Treats* of an activity or a strategy. Specifically, with respect to drones, the chapter's intent is to explain and balance all these elements objectively.

UAV: This acronym indicates an Unmanned Aerial Vehicle, and it is a synonym of *drone*. The main peculiarity of this vehicle consists of not carrying a person aboard which, as it has been outlined during the years, can be classified as an advantage as well as a disadvantage. On one hand, drones do not put the life of the operator in danger during dangerous missions; on the other hand, however, multiple criticisms have been underlined both on a legal level and on a more psychological one (and both aspects have been discussed within the chapter).

Chapter 13
Ethics and Managerial Mindset in Politics

Maria Luisa Nardi
Independent Researcher, Italy

ABSTRACT

The term strategy model appears in a variety of contexts. Though the specifics of a strategy model vary greatly depending upon the nature of the model, the basic definition of such a model remains the same in all industries and sectors, such as in policy management. Basically, a strategy model constitutes a strategic plan, or model, designed to improve a process, and it needs a specific mindset by the developer. Organizations use strategy models to improve operations and meet their goals. The development of such a model requires identifying the key goals of an organization, identifying the key components or steps of the process leading from a starting point to the achievement of those goals and creating ways of maximizing the process for optimal output and the achievement of the stated goals. Strategy models are of varying degrees of complexity.

INTRODUCTION

To date, politicians are asked above all to be able to make decisions appropriate to the problems to be faced: concluding a trade treaty, making a law to combat corruption, introducing or abolish a tax, dealing with the problem of immigration, changing or not the text of the Constitution, planning an institution or health reform. These all decisions require training in this regard, on both individual problems and identifying the unexpected effects.

DOI: 10.4018/978-1-7998-7126-2.ch013

The Policy Studies Organization was formed in 1971 in the U.S. and was also open to scholars of other nations. But the public policy as decision analysis already emerged in the 1960s, encompassing political science, sociology, economics, and other disciplines. It emerged out of greater concern for specific policy problems, the causal determinations of particular policies, and the relationship between policies and societal impact.

Historically, decision-making developed starting from some milestones: the "rational decision" theory of M. Weber (1864-1920), the "efficiency choice" theory of V. Pareto (1848-1923), the "problem orientation" theory of H. Lasswell (1902-1978) are some of the main examples.

It is necessary that every single sector policy is framed in a general political line that has characteristics of coherence and unity. That of the politician is a complex activity, which does not allow arbitrary simplifications, and requires knowledge, decision, creative design of solutions and courage.

Typical policy decision-makers are Executives, Politicians, Legislators, Bureaucrats, Policy actors, Interest groups, Research organizations, Mass media.

Following, a list of the major activities in the digital context, as an example (dall'Acqua, 2018):

- **E-Government Activities:**
    - To activate innovative projects about dematerialization, interoperability, technology infrastructures)
    - To reorganize (process reengineering) the school function diagram
    - To improve performance and efficiency with attention to quality and usefulness of the results
    - To define and implement the Regional Digital Agenda, contributing to the design and development of a Smart City
- **Open Government Activity, Practicing:**
    - Access to open information and data
    - Participation (listening and consultation), collaboration and accountability
    - Using technology as a qualifying factor for citizens' relationships and effectiveness of internal innovation processes
- **Innovative Design and Solutions:**
    - Not limited to digitizing existing document flows and processes, but
    - To take into account the opportunities for innovation that the potentialities of different application solutions can offer for knowledge management and data management (eg big data, open data)
- **Procurement of Innovative Digital Services:**
    - To activate projects and initiatives, that enrich the quality of services,

    - To rely on the support of technicians
    - To have critical capacity and interaction to identify the most effective solutions
- **Safeguard of Digital Citizenship:**
    - To ensure the respect of the principles of digital citizenship in an inclusive and disseminated way
    - To activate initiatives that make the rights of digital citizenship practicable (digital identity, privacy, and security, access to information)

Rules, norms, practices, and relationships influence individual and collective behavior. The choices of actors are explained partly by their understanding of and adherence to rules. Rules can be formal and widely understood, such as when enshrined in law or a constitution, or informal and only understood in particular organizations. Institutions at one level (e.g., constitutional) can also shape activity at another (e.g., legislative or regulatory), establish the types of venues where policy decisions are made, and make the rules that allow particular types of actors or ideas to enter the policy process.

Figure 1. Policy Cycle and activities
(freely adapted from https://debategraph.org/Details.aspx?nid=228943)

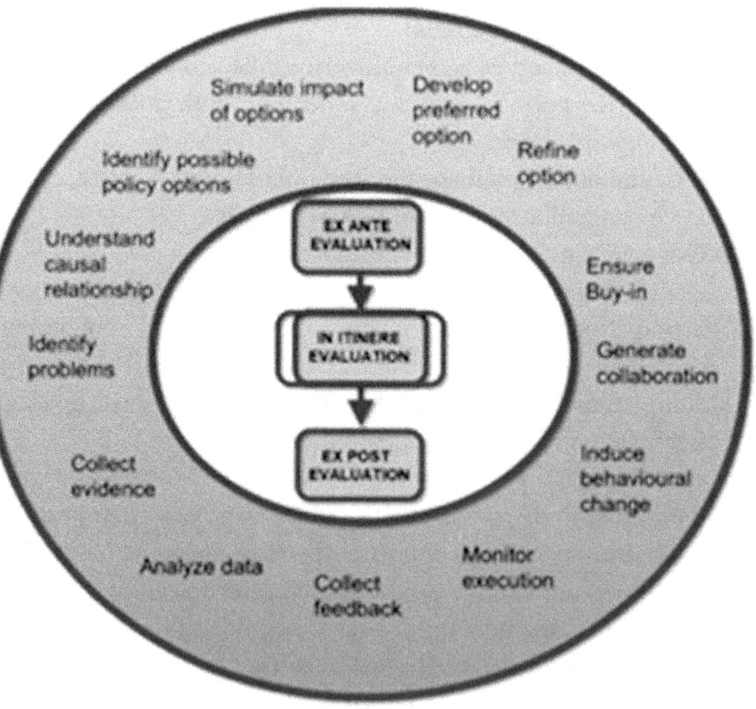

According to Selves analysis (Courtright, 2011) the factors that influence the political decision maker to be present and influential in the resolution of a crisis are:

- Very violent and rapid events that have a great media impact (e.g. earthquake)
- Events as terrorist attacks that undermine the country's security plan
- If the cause of the disaster is due to man (industrial disaster)
- Events involving state authorities at multiple levels
- Situations that can strongly influence the population (eg decision to forcibly evacuate a city area or quarantine)
- A low level of citizen satisfaction in the response put in place
- Disings in which a better work of institutional prevention would have mitigated the impact on the population
- The political color of the affected areas

These factors influence the decision on whether and how to intervene in response to emergencies both on their own territory and in support of other territories.

The foregoing does not mean that in any case a basic answer will still be given by the national emergency system. The political decision-maker in certain circumstances that we have seen above will have every interest in ensuring that the emergency response is as effective as possible to limit the loss of credibility of his Government and therefore limit the loss of votes (Chen&Healy, 2014, Strömberg, 2007).

Paul Gibbons rethinks change management with a 21st-century approach that exchanges cargo cult management for an evidence-based approach built on neuroscience and complexity sciences: "Time is overdue". And he replaces Kotter's change model with something better suited for a complex world where change is continuous and not a one-time event, and where creative change continuously drives organizational improvement (dall'Acqua, 2018): strategy models.

Strategy models offer both concrete and abstract plans for business and political management.

Concrete strategy models deal in quantifiable variables, things easily measured, analyzed, understood, and changed. Such models work directly with labor numbers, business finance, and the economy. These plans often present black and white solutions like "changing variables.

Abstract strategy models stress notions like business ethics, choice, and the availability of options.

Personal choice and beliefs of managers figure prominently in such models.

Most theories view leadership as grounded in one or more of the following three perspectives:

a) leadership as a process or relationship,
b) leadership as a combination of personality characteristics,
c) leadership as certain behaviors or, as they are more commonly referred to, leadership skills.

A few focused quotes:

"Leadership is persuading people to set aside, for a period, their individual goals, with the goal of achieving a common goal, important for the group's responsibility and well-being." (Hogan R., Curphy, Hogan J., 1994).

"Leadership helps to make the taken decisions in action. There is leadership when some people, with motivation and determined goals, mobilize resources to activate, engage and meet the motivations of others" (Leigh, 2020).

"Leadership is the quality that makes some people capable of driving others in the realization of certain activities. Leaders must achieve stability within the group, resolve conflicts, stimulate change and further improve the growth and development of the group "(Di Nubila, 2002).

A Case of Study: Orientism Processes Organization (OMP)

Following we describe one of the more interesting new interpretative paradigm, Orientism Management (OM) by Luisa dall'Acqua. The focus of this framework is specifically to orient people to manage decisions in unpredictability conditions in multi-user environment context. It intends to propose a new solution to the demand for "innovative and creative leadership" to make choices, to open to the change of perspective, to manage relationships and complex environments, basing on the conception of a new model of behavior and management of the simultaneity of how people process information.

It interprets the management activities as dynamic, adaptive and self-regulative, based not only on explicit curricular goals, but also on unpredictable social interactions and relationships between players in an emerging process of continuous change of the environment itself, to which they belong.

The model dedicates an attention for different "plans" in the *decisional process*, concerning (dall'Acqua, 2018):

- *how* do decision makers *reason* (descriptive plan) ?
- *how should* they *reason* (prescriptive plan)?
- *what* type of reasoning *motivate* the decisional action in terms of its *efficacy and efficiency* ("social" plan)?
- *what* type of reasoning is *recognizable* (communicative plan) ?

- what are *the conditions* implying a guaranteed "order" in a process, against networking and learning "chaos"?
- how can the Decision Managers be supported in their *learning of choices process and "learning path"?*
- *h*ow can Tutor be supported in own *"teaching design and teaching path* to train about decisional process?

Element of news are 5 key factors and criteria to direct and motivate people in choosing process, and following 10 different and key relationships between them. They define areas of management to improve own personal leadership and success. The concept becomes the conceptual base of an Instructional Design Model (PENTHA 2.0).

Figure 2. 10 KM typologies (OM model) with related learning modes
(Source: dall'Acqua&Md Santo, 2014)

Knowledge Management	Learning
with (Low) Consciousness	*to know what*
with (Medium) Consciousness	*to know how*
with (High) Consciousness	*to know where*
with Meaning	*to know why*
with Feeling	*to know moving*
with Will	*to know experiencing*
with Understanding	*to know enabling*
with Personalization	*to know changing*
with Availability	*to know opening*
with Synergy	*to know devoting*

Orientism proposes a new scenario of activities. to improve the people's ability to make contextual changes in natural, social, artificial multi-user environments, with multiple reference points and multi-interpreting paradigms.

Figure 3. From manager to leader, according to OM
(Source: dall'Acqua, 2018)

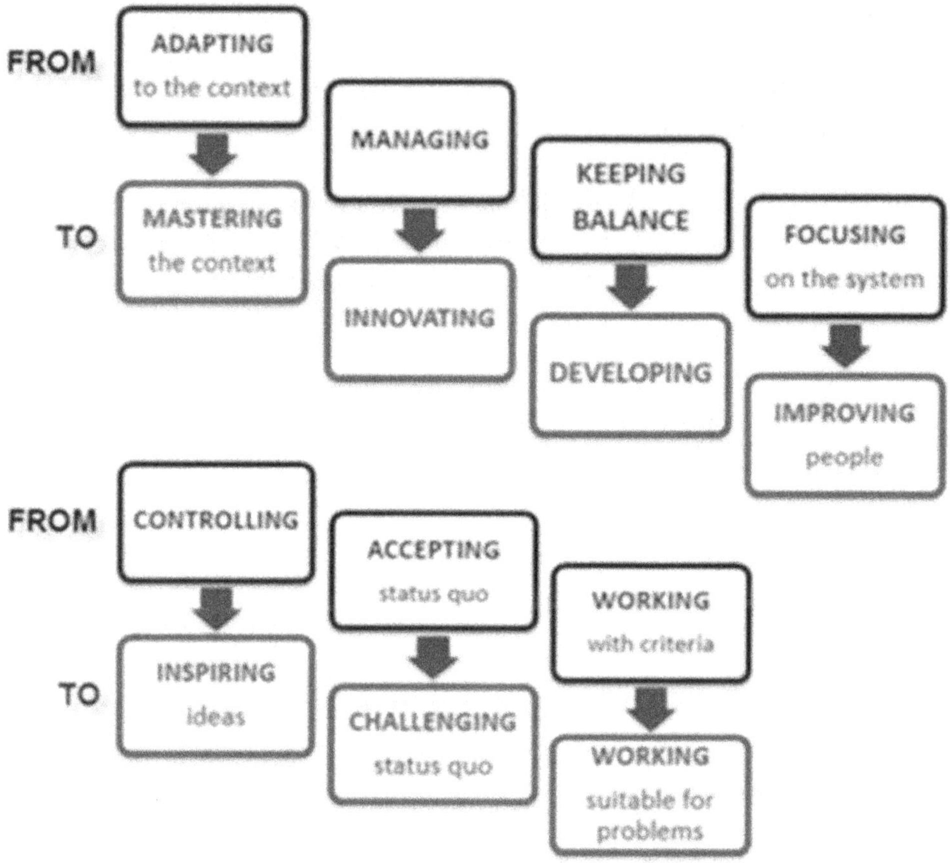

Technology's Support

State-of-the-art ICT-enabled, academic knowledge management environments provide an advanced experience where trainee and/or teamwork are lead figures, and their human and learning behaviors, cognitive preferences and human characteristics are recognized, considered, monitored, recorded, and modelled. This facilitates the capability to personalize, adapt and improve the knowledge management processes in tune with individual preferences, characteristics and preferred learning styles/habits.

It is a next-generation intelligent knowledge management environment, a mash between PLE (Personal Learning Environment), PKE (Personal Knowledge Environment), and PWE (Personal Work Environment), able to support social networking activities as well as collaborative learning, supported by an intelligent system based on artificial intelligence tools.

PLE (Personal Learning Environment) is "... a space in which subjects interact and communicate, and whose ultimate purpose is the learning and development of collective know-how. In terms of technologies used, PLEs consist of a set of loosely correlated tools, which may also include web 2.0 tools, used to work, learn, reflect and collaborate with others "(Attwell et al., 2008, p.80).

PKM (Personal Knowledge Management) is an open question about what is the PKM roles and values in different individual, organisation and social context. The main related literature by different scholars provides insight of the definition and nature of PKM. A valuable synthesis of the more significant designs is the conceptual model of PKM 2.0, developed by Cheong & Tsui. The Authors reviewed the results of a global survey about the roles and values of the main PKM frameworks (Cheong&Tsui, 2010): from mere individual activities to outcome/impact oriented; from information handling skills to personal competencies, sensemaking and self-reflection; from individually focused to a community and social collaborative focused.

P. Dorsey, of the Millikin University (US), defined seven Personal Knowledge Management (PKM) skills in (Avery et Al, 2001):

- ***Evaluating Information:*** The evaluation of information focuses on both the quality and relevance of information. Evaluation can take place as part of the retrieval process itself or as a phase engaged in after the retrieval process. The relevance issue relates to the relatedness of information to the action at hand; the quality issue relates to judgments about the accuracy of the information. Reference librarians have developed various heuristics that have been organized through their web sites to facilitate the process of evaluating and assessing information
- ***Retrieving Information:*** Relational databases, electronic library databases, Web sites, threaded discussions groups, recorded chats, and moderated and unmoderated lists are some of those information sources that have grown almost immeasurably within the last decade. Effective retrieval requires that considerable effort be placed on framing inquiry even before information retrieval commences; familiarity with subjects and keywords is central. Different search tools are based on different premises (e.g., the nature of indexing), and effective use of those tools requires some understanding of those premises. Different information resources will be useful for different types of knowledge workers, and it will be important as part of the development of knowledge workers to develop an understanding of the relative usefulness of these different information resources to support both their actions and their

personal development. Obviously, satisficing, not optimizing, is the reigning concept in retrieval.

- ***Organizing Information:*** organizing information as one of their central Personal Knowledge Management skills. They emphasize the development of coherent principles for an organization of folders to give structure to the work of the MBA student; chronological, functional, and role-based approaches have been explored. These types of organization facilitate the learning process by supporting the connection of new information to old information within the human processing system.
- ***Collaborating Around Information:*** not just within, but between organizations, defines global enterprises. Electronic mail, various forms of conferencing, and web-based structures for collaboration increasingly provide the infrastructure for the work of the enterprise and of the knowledge worker. The virtual enterprise is, in fact, defined by its collaboration around information in conjunction with the strategic purposes of the enterprise.
- ***Analyzing Information:*** is fundamental to the process of converting information into knowledge (Avery, et.al, 2001). Analysis builds on the organization of information, but goes beyond it in its emphasis on the importance of frameworks, models, and theories grounded in the standards of public communities. Analysis of information addresses the challenge of extracting meaning out of data. There are many information technology tools for analysis, but here the focus is on three: simulation software, spreadsheet software, and statistical software.
- ***Presenting Information:*** Effectiveness in presentation is critical in many areas of global business, especially marketing. An effective presentation assumes not only an understanding of audience, but a clear understanding of the purpose of the presentation as it relates to audience. The history and theory of rhetoric provides an abundant literature for guidance in the exercise of this skill. The emergence of new electronic tools and venues for presentations, through computer-based presentation tools and web sites, makes attention to this information skill even more important.
- ***Securing Information:*** has traditionally not been identified as a central information skill that needs to be developed. However, two larger developments in the global business environment conspire to make this skill central, for both organizations and knowledge workers. One of those developments is the increasing recognition of the strategic uses of information and knowledge, and the strategic benefit to be gained from sustained asymmetries in information. The second development is the explosive growth of networked environments that magnify both the risks, and the opportunities, associated with information sharing. The significant growth in intranets, which integrate organizations,

and extranets, which connect organizations with their customers and suppliers, intensifies this larger concern for securing information.

Today's always connected, instant access environment has blurred the lines between the physical office and the place where work actually happens. As the distinction between professional and personal life dissolves, and the workplace becomes truly digital, employees are communicating and collaborating in unprecedented ways. To enable knowledge sharing across the organization, they want the ability to forge productive business relationships beyond natural work groups. As a result, it is increasingly clear that the traditional 'create and push' information approach no longer meets employees' evolving needs. **Personal Work Environments** seem to be promising venues to deliver screening and brief intervention, as they potentially afford easy access to large groups of problem drinkers. Virtual work software can boost remote team's success by helping both in-house and off-site teams stay connected on projects regardless of proximity.

The **Personal Learning Enviroments** were initially presented as a paradigm shift with respect to the trainingal structures built within the platforms used for training in the institutional setting. While Learning Management Systems, used for the delivery of formal courses, were criticized because they tended to reproduce in the virtual environment the structures used in normal face-to-face teaching, with a rigid division of roles between trainers, tutors and students, of static contents predefined by the trainer, in the PLE personal and self-managed learning paths are structured, which sometimes risk leading to self-referentiality and dispersion, but which can also lead to truly personalized and motivating training courses

Each user needs to build, store and organize information and knowledge in the way that best helps them.

Human Intelligence (HI) is leveraging the life's experience, training and knowledge by engaging the user in a journey to share their life knowledge and better understand their life passion and purposes.

Users needs to interact with topic/issues/events/people/products.

Collaborative strategies are crucial, in relation with the training objectives and path, as well as the availability of adequate networking facilities for group interactions, within the virtual knowledge space, and the possibility to create communities of practice, a dynamic-, synergistic- and collaborative construction of significant knowledge. It suggests several activities, such as Community Laboratory, Group Laboratory (Project Work), Community online sessions, Community Tutoring (in synchronous and asynchronous modes), specifically assessed. The approach is implementing a specific dedicated KM platform to monitor the knowledge path.

Specific Technical Assets:

- **Content Management System:** is a software tool, installed on a web server, whose task is to facilitate the management of website content, freeing the webmaster from specific technical programming knowledge
- **Document Management System:** is a category of software systems that serves to organize and facilitate the collaborative creation of documents and other content.
- **E-learning applications:** the use of multimedia technologies and the Internet to improve the quality of learning by facilitating access to resources and services, as well as to remote exchanges and remote collaboration (in accordance to the user behaviours, role, profile, preferences and personal requirements)
- **Knowledge Objects:** A user-defined block of logic that enables to leverage information in specific ways to infer meaning from data, to interpret, classify, enrich, normalize, and model data. You can create, edit, sharing, re-use and manage knowledge objects (personal/social e-portfolio)
- **Content Ranking:** Quality content is how engaging, informing, supporting and delighting an audiences. Creating authentic, valuable content is also critical for search engine visibility. Researching the keywords (the search terms your target audience is using) is perhaps the most important factor after creating good content. It is based on Human Preference
- **Content Relevance based on User Participation and Selection**
- **Conceptual managing tool:** Knowledge work consists of non-routine, complex tasks which involve the use of large quantities of (often incomplete or ambiguous) information, both as inputs and outputs of work processes. Thus, knowledge workers need tools that add value and context to information as they work with it. These tools should reduce complexity through aggregation, organize information through categorization, and make options for action systematically visible
- **Narrative Environment:** A narrative environment is a space, physical or virtual, in which stories can unfold. A virtual narrative environment could be the narrative framework in which a story can proceed. It is argued that 'narrative environment' is not only polysemous but also paradoxical; not only representational but also performative; and not just performatively repetitive but also reflexive and constitutive. Narrative environments are therefore, sophistically, performative-representative both of the corporate dominance of life worlds and of the undoing of that dominance, through the iterative responses to the paradoxical injunction: "learn to live".
- **Semantic Search Engine:** Semantic Search is a search field that seeks to improve search accuracy in web search engines by trying to understand the

intent of the researcher and the contextual meaning of the terms used in order to generate more relevant results.

- **Gamification**
- **Intelligent Monitoring, Rule Engine based:** a "Rule-Engine" is capable of making decisions based on received data. "Expert Systems" has already been on research to address this problem.
- **Intelligent Tutoring System/Decision support (Cognitive Tutoring):** Cognitive Tutor programs give trainers the opportunity to receive individualized attention, maximizing the amount of time spent actively learning and mastering fundamental sets of knowledge skills
- **Database design to object-oriented:** An object-oriented database (OODBMS) or object database management system (ODBMS) is a database that is based on object-oriented programming (OOP). The data is represented and stored in the form of objects.
- **Online Analytical Processing (OLAP):** it designates a set of software techniques for the interactive and fast analysis of large amounts of data, which can be examined in rather complex ways.
- **Model-View-Controller based user interface:** The Model-View-Controller (MVC) architectural pattern is used in software engineering to allow for the separation of three common features: the data access (typically via a database), the business logic (how the data will be used), user interaction (how the data and actions will be visually presented)
- **Groupware system:** they are IT tools improving internal communication – between company's employees, and external communication – in relations with their clients.
- **Knowledge templates for self learning/teaching/CRM dynamic paths:** it is a step-by-step guide to complete a task
- **Video broadcasting tool** (with Chat, Raise Hand functionality and event scheduler, real-time collaborative whiteboards)
- **Social Content Viewer** to search social media (such as from Twitter, Reddit, Tumblr, YouTube, Google+, Flickr, Imgur and Wikipedia, and so on); to favorite, share and open social post and links, Social post category filtering, Hot (popular) post ordering; Use real-time social trending keywords
- **Annotations tools, Enhanced Dashboard analytics**

CONCLUSION

If we can speak of a possible difficulty run from politics in our time, it follows that, even more so, today it is a question of starting again from the examples of those who have tried to discuss and change politics even in the most difficult situations, and extrapolate a useful theoretical model, self-regulating.

Political uses of power demand explicit consideration of ethical restraints, in part because current management theory focuses on the value of outcomes rather than on the value of the means chosen. This work cited a normative model of ethical analysis that can be helpful in determining what these restraints are.

Methods and technologies are allowing to improve them. The cited framework is applied in a training environment for Personal Knowledge Management (PKM), to develop skills from management to leadership.

REFERENCES

Attwell. (2008). *Maturing Learning: Mashup Personal Learning Environments*. Academic Press.

Avery. (2001). Personal Knowledge Management (PKM): Framework for Integration and Partnerships. *Annual Conference of the Association of Small Computer Users in Education (ASCUE).*

Black, M. (1994). *The Managerial Grid*. Gulf Publishing.

Cheong, R. K. F., & Tsui, E. (2010). The Roles and Values of Personal Knowledge Management: An exploratory study. *VINE Journal of Information and Knowledge Management Systems*, *40*(2), 204–227.

Cheong, R. K. F., & Tsui, E. (2011). From Skills and Competencies to Outcome-based Collaborative Work: Tracking a Decade's Development of Personal Knowledge Management (PKM) Models. *Knowledge and Process Management. The Journal of Corporate Transformation*, *18*(3), 175–193.

Courtright, S. H. (2011). Self-Leadership: A Multilevel Review. *Journal of Management*, *37*(1), 185–222. doi:10.1177/0149206310383911

dall'Acqua, L., & Santo, M. (2012). New interpretative paradigms to understand and manage the unpredictability of the dynamics of learning in a complex multi-user knowledge environment. *Proceedings of Artificial Intelligence Workshop 2012 on Collaboration and Intelligence in Blended Learning*, 92-9.

dall'Acqua, L., & Santo, M. (2014). Orientism, the basic pedagogical approach of PENTHA ID Model vs. 2, to manage decisions in unpredictability conditions. *Proceedings of World Congress on Engineering and Computer Science WCECS2014, 1*, 316-21.

dall'Acqua, L. (2014). Needs and strategies of KM in a multi-user environment: PENTHA Model view. *Proceedings of WCS International Conference.*

dall'Acqua, L. (2018). Orientism Management Strategy for Entrepreneurial Mindset in the School Governance. In L. dall'Acqua, & D. Lukose (Eds.), *Improving Business Performance Through Effective Managerial Training Initiatives* (pp. 115-146). Hershey, PA: IGI Global.

Di Nubila, L. (2008). *Dal Gruppo al Gruppo: la formazione in team*. Pensa Multimedia.

Hogan, R., Curphy, G. J., & Hogan, J. (1994). What we know about leadership: Effectiveness and personality. *The American Psychologist, 49*(6), 493–504.

Leight, A. L. (2020). Humanistic leadership in different cultures: Defining the field by pushing boundaries. *Cross Cultural & Strategic Management, 27*(4), 533–546.

KEY TERMS AND DEFINITIONS

Computational Thinking (CT): A problem-solving process that includes a number of characteristics and dispositions. CT is essential to the development of computer applications, but it can also be used to support problem solving across all disciplines, including the humanities, math, and science.

Orientism Management (OM): A new multi-dimensional KM approach to improve the people's ability to manage decisions and own change of perspectives, according to natural, social, artificial environments, in personalized multi-user dynamic, assigned value to multiple reference points and multi-interpreting paradigms.

PKE (Personal Knowledge Environment): It is a space and tool that emphasizes learning-to-learn information skills: retrieving information, evaluating information, organizing information, collaborating around information, analyzing information, securing information and presenting information. These information skills underlie an effective strategy for integrating information technology into the curriculum rather than focusing on technology as an end in itself. The PKM framework emphasizes technology as integral to the inquiry and problem solving processes that are fundamental to both general education and academic majors.

PWE (Personal Work Environment): It is a virtual work environment can boost remote team's success by helping both in-house and off-site teams stay connected on projects regardless of proximity.

Compilation of References

discovery Official Website. (2021). Retrieved from https://www.5discovery.com/en/conflict-management/

ADB – Asia Development Bank. (n.d.). Retrieved from: http://www.adb.org

Additional Protocol to the Geneva Conventions of 12 August 1949 for the Protection of Victims of International Armed Conflicts, adopted on 8 June 1977. (n.d.). https://www.admin.ch/opc/it/classified-compilation/19770112/index.html

Agency France-Presse. (2019). US strikes in Somalia won't stop Shabaab, says US general. *Daily Nation.* https://www.nation.co.ke/news/world/US-General--Strikes-in-Somalia-won-t-stop-Al-Shabab/1068-4972234-h7vygv/index.html

Akkad, O. E. (2018). *American War: A Novel.* Vintage.

Alcañiz, M., Parra, E., & Chicchi Giglioli, I. A. (2018). Virtual Reality as an Emerging Methodology for Leadership Assessment and Training. *Frontiers in Psychology*, *9*, 1658. doi:10.3389/fpsyg.2018.01658 PMID:30258378

Ali, S. & Khan, A. A. (2017). Drone strikes, Logic of Self-defense and Violation of State Sovereignty. *Pakistan Journal of History and Culture, 38*(2), 187-206.

Alim, A. (2018). International Armed Conflict and Drone Warfare: Quest for a New Apace in international Humanitarian Law. *International Journal of Law, Humanities & Social Science, 2*(4), 70-82.

Alley, R. (2013). *The Drone Debate, Sudden Bullet or Slow Boomerang?* Discussion Paper in Centre for Strategic Studies: New Zealand, Victoria University of Wellington.

Amazon.com. (n.d.). https://www.amazon.com/Amazon-Prime-Air/b?ie=UTF8&node=8037720011

American Congress. (2001). *Public Law 107–40—SEPT. 18, 2001, 107th Congress, Authorization for Use of Military Force. 50 USC 1541 note.*

Amidon, M. (2005). *Groupthink, politics, and the decision to attempt the Son Tay rescue.* Air War Coll Maxwell AFB AL.

Amnesty International. (2013). *"Will I be next?" US Drone strikes in Pakistan.* Amnesty International Publications.

Annual Disaster Statistic Review from the Emergency Events Database EM-DAT. (2011). *Centre for Research on the Epidemiology of Disasters (CRED).* Available at https://www.emdat.be/publications

ASEAN – Association of South East Asian Nations. (n.d.). Retrieved from: http://www.asean.org

Asher, T., Ogle, E., Bailenson, J. N., & Herrera, F. (2018). *Becoming homeless: a human experience. ACM SIGGRAPH 2018 Virtual, Augmented, and Mixed Reality.* doi:10.1145/3226552.3226576

Attenborough, D. (2020). *A Life on Our Planet.* Grand Central Publishing.

Attwell. (2008). *Maturing Learning: Mashup Personal Learning Environments.* Academic Press.

AU – African Union. (n.d.). Retrieved from: http://www.au.int

Avery. (2001). Personal Knowledge Management (PKM): Framework for Integration and Partnerships. *Annual Conference of the Association of Small Computer Users in Education (ASCUE).*

Baharmand, H., Comes, T., & Lauras, M. (2017). Managing in-country transportation risks in humanitarian supply chains by logistics service providers: Insights from the 2015 Nepal earthquake. *International Journal of Disaster Risk Reduction, 24*, 549–559. doi:10.1016/j.ijdrr.2017.07.007

Bandler, R. (1996). DHE. Meta Pubns.

Bandler, R., Robert, A., & Fitzpatrick, O. (2013). *An Introduction To Nlp.* Harper Collins Publishers.

Barber, J. (2018). *Imagining Sustainable Futures in Popular Culture: Moving Beyond Dystopia and Techno-fantasy Narratives.* IAMCR 2018 Conference, Eugene, OR.

BeckwithP. (2020). *Home Page.* https://paulbeckwith.net/

BensonK. (2014). "Kill 'em and Sort it Out Later": Signature Drone Strikes and International Humanitarian Law. *Pacific McGeorge Global Business & Development Law Journal, 27*(1). https://ssrn.com/abstract=2465620

Bergen, P., & Rowland, J. (2012). *Obama ramps up covert war in Yemen.* CNN. https://edition.cnn.com/2012/06/11/opinion/bergen-yemen-drone-war/index.html

Bergen, P., Sterman, D., Salyk-Virk, M., Sims, A., & Ford, A. (n.d.a). *The War in Yemen.* New America. https://www.newamerica.org/in-depth/americas-counterterrorism-wars/us-targeted-killing-program-yemen/

Bergen, P., Sterman, D., Salyk-Virk, M., Sims, A., & Ford, A. (n.d.b). *The War in Somalia.* New America. https://www.newamerica.org/international-security/reports/americas-counterterrorism-wars/the-war-in-somalia/

Bergen, P., Sterman, D., Salyk-Virk, M., Sims, A., & Ford, A. (n.d.c). *The Drone War in Pakistan*. New America. https://www.newamerica.org/in-depth/americas-counterterrorism-wars/pakistan/

Birkie, S. E., & Trucco, P. (2020). Do not expect others do what you should! Supply chain complexity and mitigation of the ripple effect of disruptions. *International Journal of Logistics Management, 36*(7), 27–29.

Bituere, J. (2020). *Estudo sobre a Otimização da Cadeia de Abastecimento de "Marine Gasoil" na Marinha Portuguesa* [MSc Dissertation]. Portuguese Naval Academy. (in Portuguese)

Black & Moudon. (1994). *The Managerial Grid*. Gulf Publishing.

Blair, O. (2021, January 27). 125 coronavirus memes to get you through self-isolation and social distancing. *Elle*. https://bit.ly/3gkvblr

Bode, C., Wagner, S. M., Petersen, K. J., & Ellram, L. M. (2011). Understanding responses to supply chain disruptions: Insights from information processing and resource dependence perspectives. *Academy of Management Journal, 54*(4), 833–856. doi:10.5465/amj.2011.64870145

Bourdieu, P. (1986). Habitus, code et codification. *Actes de la Recherche en Sciences Sociales, 64*(1), 40–44. doi:10.3406/arss.1986.2335

Bowden, M. (2013). The Killing Machines, How to think about drones. *The Atlantic*. https://www.theatlantic.com/magazine/archive/2013/09/the-killing-machines-how-to-think-about-drones/309434/

Boyle, M. J. (2015). The legal and ethical implications of drone warfare. *International Journal of Human Rights, 19*(2), 105–126. doi:10.1080/13642987.2014.991210

Brooks, D. (2013). Florence and the Drones. *The New York Times*. https://www.nytimes.com/2013/02/08/opinion/brooks-florence-and-the-drones.html

Brooks, R. (2013). *10 ways to fix the drone war*. Foreign Policy. https://foreignpolicy.com/2013/04/11/10-ways-to-fix-the-drone-war/

Brooks, R., & Hastings, M. (2012). The Rise of the Killer Drones: How America Goes to War in Secret. *Rolling Stone*. https://www.rollingstone.com/politics/politics-news/the-rise-of-the-killer-drones-how-america-goes-to-war-in-secret-231297/

Brooks, S. K., Webster, R. K., Smith, L. E., Woodland, L., Wessely, S., Greenberg, N., & Rubin, G. J. (2020). The psychological impact of quarantine and how to reduce it: A rapid review of the evidence. *Lancet, 395*(10227), 912–920. Advance online publication. doi:10.1016/S0140-6736(20)30460-8 PMID:32112714

Brown, G. (2018). *Dirt to Soil, One family's journey into regenerative agriculture*. Chelsea Green Publishing.

Bruguera, C., Guitert, M., & Romeu, T. (2019). Social media and professional development: A systematic review. *Research in Learning Technology, 27*(0). Advance online publication. doi:10.25304/rlt.v27.2286

Bruner, J. (1986). *Actual minds, possible worlds.* Harvard University Press. doi:10.4159/9780674029019

Brunstetter, D., & Braun, M. (2011). The implications of drones on the just war tradition. *Ethics & International Affairs, 25*(3). https://www.cambridge.org/core/journals/ethics-and-international-affairs/article/abs/implications-of-drones-on-the-just-war-tradition/97ABF476B8494CC44A71E011DD8B7600

Burkov, A. (2019). *The Hundred-Page Machine Learning Book.* Andriy Burkov.

Bush, G. W. (2001). *Address Before a Joint Session of the Congress on the United States Response to the Terrorist Attacks of September 11.* The American Presidency Project. https://www.presidency.ucsb.edu/documents/address-before-joint-session-the-congress-the-united-states-response-the-terrorist-attacks

Business Insider Intelligence. (2020). *Drone market outlook: industry growth trends, market stats and forecast.* Business Insider. https://www.businessinsider.com/drone-industry-analysis-market-trends-growth-forecasts?IR=T

Byman, D. L. (2013). *Why Drones Work: The Case for Washington's Weapon of Choice.* Brookings. https://www.brookings.edu/articles/why-drones-work-the-case-for-washingtons-weapon-of-choice/

Byrne, E. F. (2015). Making Drones to Kill Civilians: Is it Ethical? *Journal of Business Ethics*, 1-13. . doi:10.100710551-015-2950-4

Cambridge Dictionary. (n.d.). https://dictionary.cambridge.org/it/dizionario/inglese/uav

Campbell, J. (1991). *The Power of Myth, with Bill Moyers.* Anchor Books, Random House.

Campbell, J. (2008). *The Hero with a Thousand Faces.* New World Library. (Original work published 1947)

Campbell, J., & Moyers, B. (1988). *The Power of Myth [DVD].* Kino Lorber.

CARICOM – Caribbean Community for Regional Integration. (n.d.). Retrieved from: http://www.caricom.org

Carruthers, S. L. (2009). *Cold war captives: Imprisonment, escape, and brainwashing.* University of California Press. doi:10.1525/9780520944794

Casey-Maslen, S. (2012). Pandora's box? Drone strikes under jus ad bellum, jus in bello, and international human rights law. *International Review of the Red Cross, 94*(886), 597–625. doi:10.1017/S1816383113000118

Cavallaro, J., Sonnenberg, S., & Knuckey, S. (2012). Living Under Drones: Death, Injury and Trauma to Civilians from US Drone Practices in Pakistan. Stanford: International Human Rights and Conflict Resolution Clinic, Stanford Law School; New York: NYU School of Law, Global Justice Clinic.

Cavoukian, A., & Office of the Information and Privacy Commissioner/Ontario. (2012). *Privacy and drones: Unmanned aerial vehicles*. Ontario, Canada: Information and Privacy Commissioner of Ontario, Canada.

Chamayou, G. (2015). *A Theory of the Drone*. The New Press.

Charteris, J., Thomas, E., & Masters, Y. (2018). Funds of identity in education: Acknowledging the life experiences of first-year tertiary students. *Teacher Educator, 53*(1), 6–20. doi:10.1080/08878730.2017.1367057

Chase, R. & Jacobs, F.R. (2017). *Operations and Supply Chain Management.* New York: McGraw-Hill Ed.

Chen, J., Chen, T. H. Y., Vertinsky, I., Yumagulova, L., & Park, C. (2013). Public-private partnerships for the development of disaster resilient communities. *Journal of Contingencies and Crisis Management, 21*(3), 130–143. doi:10.1111/1468-5973.12021

Cheong, R.K.F., & Tsui, E. (2011). From Skills and Competencies to Outcome-based Collaborative Work: Tracking a Decade's Development of Personal Knowledge Management (PKM) Models. *Knowledge and Process Management. The Journal of Corporate transformation, 18*(3), 175-193.

Cheong, R. K. F., & Tsui, E. (2010). The Roles and Values of Personal Knowledge Management: An exploratory study. *VINE Journal of Information and Knowledge Management Systems, 40*(2), 204–227.

Cheong, R. K. F., & Tsui, E. (2011). From Skills and Competencies to Outcome-based Collaborative Work: Tracking a Decade's Development of Personal Knowledge Management (PKM) Models. *Knowledge and Process Management. The Journal of Corporate Transformation, 18*(3), 175–193.

Chopra, S., & Sodhi, M.S. (2004). Managing risk to avoid supply chain breakdown. *Massachusetts Institute of Technology Sloan Management Review, 46*(1), 53-61.

Christakis, C., & Sean, B. (2016). The Serengeti Rules, The quest to discover how life works and why it matters. Princeton University Press.

Christopher, M., & Lee, H. (2004). Mitigating supply chain risk through improved confidence. *International Journal of Physical Distribution & Logistics Management, 34*(5), 388–396. doi:10.1108/09600030410545436

Clark, K. (2017). *Drone warfare 1: Afghanistan, birthplace of the armed drone*. Afghanistan analysts Network. https://www.afghanistan-analysts.org/en/reports/international-engagement/drone-warfare-1-afghanistan-birthplace-of-the-armed-drone/

Clarke, A. C. (2001). *A Space Odyssey. Hutchinson*. UK New American Library.

Cloud, D. S. (2010). CIA drones have broader list of targets. *Los Angeles Times*. https://www.latimes.com/world/la-xpm-2010-may-05-la-fg-drone-targets-20100506-story.html

CNN. (2003). *'Master' and 'slave' computer labels unacceptable, officials say*. CNN.

Cooke, N. J., Pringle, H., Pedersen, H., & Connor, O. (2006). *Human Factors of Remotely Piloted Vehicles. Advances in Human Performance and Cognitive Engineering Research Series*. Elsevier.

Courtright, S. H. (2011). Self-Leadership: A Multilevel Review. *Journal of Management, 37*(1), 185–222. doi:10.1177/0149206310383911

Cozzolino A. (2014). Agilità nella logistica delle emergenze. Le imprese apprendono dalle organizzazioni umanitarie. *Sinergie, Italian Journal of Management, 95*, 75-98.

Cozzolino, A. (2013). *L'agilità nella logistica delle emergenze. Come le imprese apprendono la gestione dell'imprevedibile dalle organizzazioni umanitarie nelle operazioni di primo soccorso*. Convegno Aidea, Lecce, "Il ruolo dell'azienda nell'economia. Esiste un modello aziendale orientato alla crescita?

Cozzolino, A., Rossi, S., & Conforti, A. (2012). Agile and lean principles in the humanitarian supply chain. The case of the United Nations World Food Programme. *Journal of Humanitarian Logistics and Supply Chain Management, 2*(1), 16-33.

Cozzolino, A. (2012). *Humanitarian logistics. Cross-sector cooperation in disaster relief management*. Springer Briefs in Business Series. Springer.

Cronin, A. K. (2013). *Why Drones Fail: When Tactics Drive Strategy*. Foreign Affairs. https://www.foreignaffairs.com/articles/somalia/2013-06-11/why-drones-fail

Crosston, M. (2014). Pandora's Presumption: Drones and the Problematic Etchis of Techno-War. *Journal of Strategic Security*, *7*(4), 1–24. doi:10.5038/1944-0472.7.4.2

Currier, C., & Maass, P. (2015). *Firing blind: Flawed intelligence and the limits of drone technology: The drone papers*. The intercept. https://theintercept.com/drone-papers/firing-blind/

dall'Acqua, L. (2014). *Needs and strategies of KM in a multi-user environment: PENTHA Model view and analysis*. In Jekpot KM19, Milano, Italy.

dall'Acqua, L. (2014). Needs and strategies of KM in a multi-user environment: PENTHA Model view. *Proceedings of WCS International Conference*.

dall'Acqua, L. (2018). *Improving Business Performance Through Effective Managerial Training Initiatives*. IGI Global.

dall'Acqua, L., & Gironacci, I.M. (2020). *Political Decision-Making and Security Intelligence: Recent Techniques and Technological Developments*. Hershey, PA: IGI Global.

dall'Acqua, L., & Santo, M. (2012). New interpretative paradigms to understand and manage the unpredictability of the dynamics of learning in a complex multi-user knowledge environment. *Proceedings of Artificial Intelligence Workshop 2012 on Collaboration and Intelligence in Blended Learning*, 92-9.

dall'Acqua, L., & Santo, M. (2012). New interpretative paradigms to understand and manage the unpredictability of the dynamics of learning in a complex multi-user knowledge environment. *Proceedings of Artificial Intelligence Workshop 2012 on Collaboration and Intelligence in Blended Learning.*

dall'Acqua, L., & Santo, M. (2014). Orientism, the basic pedagogical approach of PENTHA ID Model vs. 2, to manage decisions in unpredictability conditions. *Proceedings of World Congress on Engineering and Computer Science WCECS2014, 1*, 316-21.

dall'Acqua, L., & Santo, M. (2014). Orientism, the basic pedagogical approach of PENTHA ID Model vs. 2, to manage decisions in unpredictability conditions. *Proceedings of World Congress on Engineering and Computer Science WCECS2014*, 316-21.

dall'Acqua, L. (2018). Orientism Management Strategy for Entrepreneurial Mindset in the School Governance. In L. dall'Acqua, & D. Lukose (Eds.), *Improving Business Performance Through Effective Managerial Training Initiatives* (pp. 115-146). Hershey, PA: IGI Global.

Davis, G. H. (1977). Prisoners of war in twentieth-century war economies. *Journal of Contemporary History, 12*(4), 623–634. doi:10.1177/002200947701200401

Dawkins, R. (2004). *The Ancestor's Tale, The dawn of evolution*. Houghton Mifflin.

De Vignemont, F., & Singer, T. (2006). The empathic brain: How, when and why? *Trends in Cognitive Sciences, 10*(10), 435–441. doi:10.1016/j.tics.2006.08.008 PMID:16949331

Decety, J. (2010). The neurodevelopment of empathy in humans. *Developmental Neuroscience, 32*(4), 257–267. doi:10.1159/000317771 PMID:20805682

Decety, J., & Meyer, M. (2008). From emotion resonance to empathic understanding: A social developmental neuroscience account. *Development and Psychopathology, 20*(4), 1053–1080. doi:10.1017/S0954579408000503 PMID:18838031

Denning, P., & Tedre, M. (2019). *Computational Thinking*. The MIT Press Essential Knowledge Series.

Di Loreto, M. (2006). *Impiego Joint e Combined di Unmanned Aerial Vehicles (UAV): Stato dell'arte e prospettive future di impiego*. eMiSS, Centro Militare di Studi Strategici.

Di Nubila, L. (2008). *Dal Gruppo al Gruppo: la formazione in team*. Pensa Multimedia.

Duvenage, M. A. (2010). *Intelligence Analysis in the Knowledge Age-An Analysis of the Challenges facing the Practice of Intelligence Analysis*. Stellenbosch University.

Edwards, A. (Ed.). (2017). *Working relationally in and across practices: A cultural-historical approach to collaboration.* Cambridge University Press. doi:10.1017/9781316275184

Ellison, N. B., Steinfield, C., & Lampe, C. (2011). Connection strategies: Social capital implications of Facebook-enabled communication practices. *New Media & Society*, *13*(6), 873–892. doi:10.1177/1461444810385389

Englander, J. (2013). *High Tide on Main Street, Rising sea level and the coming coastal crisis.* The Science Bookshelf.

Ericsson Annual Report. (2019). *Building a stronger company long term.* Author.

Ericsson Response Fact Sheet. (2020). *Breaking the energy curve.* Author.

EU – European Union. (n.d.). Retrieved from: http://www.europa.eu

Feroz, E. (2020). *Death by drone: America's vicious legacy in Afghanistan.* Foreign Policy. https://foreignpolicy.com/2020/03/27/afghanistan-drones-america-legacy-taliban/

Fischer, D. (2014). *Scare or care: the War on Terror 2001-2013.* Ruprecht-Karls-Universität Heidelberg, Heidelberg Center for American Studies.

Fischetti, M. (2019, August). A New Reality, Climate Change is dramatically altering life at the top of the world. *Scientific American, 321*(2), 37.

For artificial intelligence pioneer Marvin Minsky, computers have soul. (2014, May 13). *Jerusalem Post.*

Forrester, J. W. (1961). *Industrial Dynamics.* Pegasus Communications.

Forrester, J. W. (1968). *Principles of Systems.* Pegasus Communications.

Forrester, J. W. (1971). *World Dynamics.* Wright-Allen Press.

Franz, N., & Ladewig, R. (2017). Synthetic Reality and Blind Fight: The Operationalization of Perception. *Politik, 20*(1), 30–44.

Gaggioli, G. (2013). *The use of force in armed conflicts, Interplay between the conduct of hostilities and law enforcement of paradigms.* International Committee of the Red Cross' Expert Meeting.

Gettinger, D. (2020). The Soleimani killing and 5 things to know about drones in Iraq. *Bulletin of the Atomic Scientists.* https://thebulletin.org/2020/01/soleimani-and-beyond-5-ways-that-drones-have-destabilized-iraq/#

Ginestie, J., Impedovo, M., (2020). International Teachers Professional Developing: Blended Learning between Europe and Asia. *Voprosy obrazovaniya, 2,* 114–127.

Glue Official Website. (2019). Retrieved from https://glue.work/

Gober, S. J. (2005). *The reintegration of military families following long term separation.* Arizona State Univ Tempe. doi:10.1037/e452872006-001

Gore, A. (2006). *An Inconvenient Truth: The planetary emergency of global warming and what we can do about it.* New York: Viking.

Gould, S. J. (1989). *Wonderful Life, The Burgess Shale and the Nature of History*. W. W. Norton.

Graae, A. I., & Maurer, K. (2017, March). Introduction: Debating Drones: Politics, Media, and Aesthetics. *Politik, 20*(1). Advance online publication. doi:10.7146/politik.v20i1.27640

Greenwald, G. (2012). Obama moves to make the War on Terror permanent. *The Guardian*. https://www.theguardian.com/commentisfree/2012/oct/24/obama-terrorism-kill-list

Gregory, D. (2014). *Drone geographies*. Radical Philosophy. https://www.radicalphilosophy.com/article/drone-geographies

Gregory, T. A. (2013). *Drones: Mapping the Legal Debate*. Università di Auckland.

Grimm, D. (2013). Whole community planning: Building resiliency at the local level. *Journal of Business Continuity & Emergency Planning*, *7*, 403–438. PMID:24578027

GSMA – Global System for Mobile Communications Association. (n.d.). Retrieved from: http://www.gsma.com

Gusterson, H. (2017). *Drone warfare*. American Institute of Physics. https://aip.scitation.org/doi/abs/10.1063/1.5009234 doi:10.1063/1.5009234

Haas, P. (1992). Introduction: epistemic communities and international policy coordination. *International Organization. Cambridge Journals, 46*(1), 1–35. https://www.jstor.org/stable/2706951

Hambling, D. (2019). *Autonomous killer drones set to be used by Turkey against Syria*. New Scientist. https://www.newscientist.com/article/2217171-autonomous-killer-drones-set-to-be-used-by-turkey-in-syria/

Hansen, J. (2020). *Home Page*. Retrieved on October 2020, https://csas.earth.columbia.edu/

Hansen, S. J. (2019). *Somalia drone strikes are a potent weapon, but not the game changer*. The Conversation. https://theconversation.com/somalia-drone-strikes-are-a-potent-weapon-but-not-the-game-changer-111845

Hansen, J. (2011). *Storms of My Grandchildren, The truth about the coming climate catastrophe and our last chance to save humanity*. Bloomsbury Publishing.

Haraway, D. (2016). Staying with the Trouble: Making Kin in the Chthulucene. Duke University Press. doi:10.1215/9780822373780

Harvard Business Review. (2001). *Risk Management in a Time of Global Uncertainty*. Author.

Hassanalian, M., & Abdelkefi, A. (2017). Classifications, applications, and design challenges of drones: A review. *Progress in Aerospace Sciences*, *91*, 99–131. doi:10.1016/j.paerosci.2017.04.003

Hastings, M. (2012). *The Rise of the Killer Drones: How America Goes to War in Secret*. Rolling Stone. https://www.rollingstone.com/politics/politics-news/the-rise-of-the-killer-drones-how-america-goes-to-war-in-secret-231297/

Hauser, L.M. (2003). Risk-adjusted supply chain management. *Supply Chain Management Review, 7*(6), 64-71.

Hausman, W. H., Hau, L., & Subramanian, U. (2005). *Global logistics indicators, supply chain metrics and bilateral trade partners.* World Bank Policy Research Working Paper No. 3773. doi:10.2139srn.869999

Hawken, P. (2017). *Drawdown, The most comprehensive plan ever proposed to reverse global warming*. Penguin Books.

Héder, M., & Paksi, D. (2018). Non-Human Knowledge According to Michael Polanyi. *Tradition and Discovery*, *44*(1), 50–66.

Herbane, B., Elliott, D., & Swartz, E. M. (2010). Business continuity management, second edition: A crisis management approach. New York, NY: Routledge.

Hersey & Blanchard. (1982). *Management of Organizational Behavior: Leading Human Resource*. Prentice-Hall.

Hickman, L., & Akdere, M. (2017). Exploring Virtual Reality for Developing Soft-Skills in STEM Education. *2017 7th World Engineering Education Forum (WEEF),* 461-465. 10.1109/WEEF.2017.8467037

Hill, C. M. (2017). More-than-reflective practice: Becoming a diffractive practitioner. *Teacher Learning and Professional Development*, *2*(1), 1–17.

History.com Editors. (2019). *A Timeline of the U.S.-Led War on Terror*. A&E Television Networks. https://www.history.com/topics/21st-century/war-on-terror-timeline

Hoffman, F., & Kalikow, E. (2013). *To drone or not to drone*. Foreign Policy Research Institute. https://www.fpri.org/article/2013/08/to-drone-or-not-to-drone/

Hogan, R., Curphy, G. J., & Hogan, J. (1994). What we know about leadership: Effectiveness and personality. *The American Psychologist*, *49*(6), 493–504.

Homer. (2015). The Iliad (C. Alexander, Trans.). CCO HarperCollins.

Horton, T. (2017). *High Tide in Dorchester, A Bay Journal Documentary*. Retrieved from https://hightidedorchester.org/

Howard, M., & Paret, P. (1984). *Carl von Clausewitz On war*. Princeton University Press.

Hunter, M. (2017). *The End We Start From.* Grove Press.

Hussain, M. (2015). *Former drone operators say they were "horrified" by cruelty of assassination program.* The Intercept. https://theintercept.com/2015/11/19/former-drone-operators-say-they-were-horrified-by-cruelty-of-assassination-program/

Hussien, A., Tucker, M., Cotgrave, A., Al-Khafajiy, M., & Shamsa, T. B. (2019). Optimizing Project Delivery through Augmented Reality and Agile Methodologies. *2019 12th International Conference on Developments in eSystems Engineering (DeSE),* 1006-1013, 10.1109/DeSE.2019.00187

ICJ – International Court of Justice. (n.d.). Retrieved from: http://www.icj-cij.org

Ilha A. (2011). The humanitarian relief chain. *South East European Journal of Economics & Business, 6*(2), 45-54.

IMF – International Monetary Fund. (n.d.). Retrieved from: http://www.imf.org

Impedovo, M. A., & Khatoon, S. Kinley, (2019). Global South teacher educators in the digital landscape: implications for professional learning. *Research on Education and Media, 11*(2). doi:10.2478/rem-2019-0018

International Committee of the Red Cross. (2008). *How is the Term "Armed Conflict" Defined in International Humanitarian Law?* https://www.icrc.org/en/doc/resources/documents/article/other/armed-conflict-article-170308.htm#:~:text=International%20humanitarian%20law%20distinguishes%20two,or%20between%20such%20groups%20only.

International Committee of the Red Cross. (2010a). *IHL and other legal regimes.* ICRC. https://www.icrc.org/en/document/ihl-and-other-legal-regimes

International Committee of the Red Cross. (2010b). *Protection of the civilian population.* https://www.icrc.org/en/doc/what-we-do/protecting-civilians/overview-protection-civilian-population.htm

International Organization for Standardization (ISO). (2009). *Risk management – principles and guidelines.* ISO 31000:2009. Available at https://www.iso.org/obp/ui/#iso:std:iso:iwa:31:ed-1:v1:en

International Organization for Standardization (ISO). (2012). *Societal security – business continuity management systems – requirements.* ISO 22301:2012. Available at https://www.iso.org/obp/ui/#iso:std:iso:22301:ed-1:v2:en

Internet World Stats. (2020). https://www.internetworldstats.com/stats.htm

Irr, C. (2017). *Climate Fiction in English.* Retrieved from http://www.academica.edu/35288142/Climate_Fiction_In_English

Ishida, S., Montshiwa, A. L., & Nagahira, A. (2016). Modifying business continuity plan (BCP) towards an effective auto-mobile business continuity management (BCM): A quantitative approach. *Journal of Disaster Research, 11*(4), 691–698. doi:10.20965/jdr.2016.p0691

ITU – International Telecommunications Union. (n.d.). Retrieved from: http://www.itu.int

Jackson, S. (2010). *Churchill's unexpected guests: Prisoners of war in britain in world war II*. The History Press.

Jacobsen, R. (2019). Beyond Seawalls, Fortified wetlands can protect shorelines better than hard structures. *Scientific American, 32054*, 32.

Jet Propulsion Laboratory, Physical Oceanographic Distributed Active Archive Center. (2020). *Seal Level*. Author.

Kanter, R. M. (1999). From spare change to real change. The social sector as beta site for business innovation. *Harvard Business Review*, *77*(3), 122–132. PMID:10387574

Keene, S. D. (2015). *Lethal and legal? The ethics of drone strikes*. Strategic Studies Institute and U.S. Army War College Press.

Kellener, J. D. (2019). *Deep Learning*. MIT Press. doi:10.7551/mitpress/11171.001.0001

Kennedy, C., & Moore, J. (2010). *Military neuropsychology*. Springer Publishing Company.

Kidder, T. (1981). *The Soul of a New Machine*. Little, Brown and Company.

Kleindorfer, P. R., & Saad, G. H. (2005). Managing disruption risks in supply chains. *Production and Operations Management*, *14*(1), 53–68. doi:10.1111/j.1937-5956.2005.tb00009.x

Koh, H. H. (2010). *Annual Meeting of the American Society of International Law*. https://2009-2017.state.gov/s/l/releases/remarks/139119.htm

Kotter, J. P. (2012). *Leading Change*. Harvard Business School Press.

Kwak, D. W., Rodrigues, V. S., Mason, R., Pettit, S., & Beresford, A. (2018). Risk interaction identification in international supply chain logistics: Developing a holistic model. *International Journal of Operations & Production Management*, *38*(2), 372–389. doi:10.1108/IJOPM-03-2016-0121

Laineste, L., & Voolaid, P. (2017). Laughing across borders: Intertextuality of internet memes. *The European Journal of Humour Research*, *4*(4), 26–49. doi:10.7592/EJHR2016.4.4.laineste

Lee, D. (2019). *Amazon to deliver by drone 'within months'*. BBC. https://www.bbc.com/news/technology-48536319

Lee, H. (2004). The triple-A supply chain. Harvard Business Review on Supply Chain Management, 82(10), 102-112.

Lee-Morrison, L. (2014). *State of the Art Report on Drone-based Warfare*. Division of Art History and Visual Studies, Department of Arts and Cultural Sciences, Lund University.

Leight, A. L. (2020). Humanistic leadership in different cultures: Defining the field by pushing boundaries. *Cross Cultural & Strategic Management*, *27*(4), 533–546.

Lewis, M. W. (2012). Drones and the boundaries of the battlefield. *Texas International Law Journal*, *47*(2), 293–314.

Lister, T., & Cruickshank, P. (2011). *Anwar al-Awlaki: al Qaeda's rock star no more*. CNN. https://edition.cnn.com/2011/09/30/world/meast/analysis-anwar-al-awlaki/index.html

Literat, I., & van den Berg, S. (2019). Buy memes low, sell memes high: Vernacular criticism and collective negotiations of value on Reddit's Meme Economy. *Information Communication and Society*, *22*(2), 232–249. doi:10.1080/1369118X.2017.1366540

Logistics Emergency Team (LET). (2019). *Annual Report*. Available at https://logcluster.org/document/let-annual-report-2019

Long, D., & Wood, D. (1995). The logistics of famine relief. *Journal of Business Logistics*, *16*(1), 213–229.

Maak, T. (2007). Responsible leadership, stakeholder engagement, and the emergence of social capital. *Journal of Business Ethics*, *74*(4), 329–343. doi:10.100710551-007-9510-5

MacDonald, S. (2020). What do you (really) meme? Pandemic memes as social, political repositories. *Leisure Sciences*. Advance online publication. doi:10.1080/01490400.2020.1773995

Maclean, N. (2017). *Young Men and Fire*. University of Chicago Press. doi:10.7208/chicago/9780226450490.001.0001

Maggio, F. F. (2018). *Mondo Digitale: Il computer tra realtà e fantasia (Computer between reality and phantasy)*. Associazione Italiana Informatica e Calcolo Automatico.

Markets And Markets. (2018). *Military Drones Market by Type, Application (ISRT, Delivery and Transportation, Combat Operations, Battle Damage Management), Range (VLOS, EVLOS, BLOS), Propulsion Type, Launching Mode, Endurance, MTOW, Region - Global Forecast to 2025*. Markets and Markets. https://www.marketsandmarkets.com/Market-Reports/military-drone-market-221577711.html

Masters, J. (2011). *Closing Guantanamo?* Council on Foreign Relations. https://www.cfr.org/backgrounder/closing-guantanamo

Matějka, S. (2014). *Unmanned Aerial Vehicles in US National Security Policy, New Face of War on Terror*. Univerzita Karlova V Praze, Fakulta Sociálních Věd, Institut Mezinárodních Studií, Katedra Amerických Studií, Praga.

Maurer, P. (2013). *The use of armed drones must comply with laws*. International Committee of the Red Cross. https://www.icrc.org/en/doc/resources/documents/interview/2013/05-10-drone-weapons-ihl.htm

Mazzetti, M., & Sanger, D. E. (2009). *Obama expands missile strikes inside Pakistan*. The New York Times. https://www.nytimes.com/2009/02/21/washington/21policy.html

McCarley, J. S., & Wickens, C. D. (2005). *Human Factors Implications of UAVs in the National Airspace*. Institute of Aviation, Aviation Human Factors Division, University of Illinois at Urbana-Champaign.

McDonagle, J., & Vella, G. (1996). *New Archetype for Competitive Intelligence*. Praeger.

McDonnell, T. M. (2012). Sow What You Reap? Using Predator and Reaper Drones to Carry Out Assassinations or Targeted Killings of Suspected Islamic Terrorists. *The George Washington International Law Review, 44*(243), 243-316. https://digitalcommons.pace.edu/lawfaculty/859/

McDowell, D. (2008). *Strategic Intelligence*. Academic Press.

McKenna, A. (2016). The public acceptance challenge and its implications for the developing civil drone industry. In The future of drone use: opportunities and threats from ethical and legal perspectives. T.M.C. Asser Press. doi:10.1007/978-94-6265-132-6_18

McKight, B., & Linnenluecke, M.K. (2016). How firm responses to natural disasters strengthen community resilience: A stakeholder-based perspective. *Organization & Environment, 29*(3), 290-307.

McLeod, H. (2018). *Virtual Reality in Leadership Development*. Retrieved from https://www.aacsb.edu/blog/2018/august/virtual-reality-in-leadership-development

McManus, D. (2009). *The cost of killing by remote control*. Los Angeles Times. https://www.latimes.com/archives/la-xpm-2009-may-03-oe-mcmanus3-story.html

Mcowan, P. W., & Curzon, P. (2017). *Power Of Computational Thinking, The: Games, Magic And Puzzles To Help You Become A Computational Thinker*. World Scientific Publishing Europe Ltd.

Meadows, D. H. (2008). *Thinking in Systems: A Primer*. Chelsea Green Publishing Co.

Melzer, N. (2008). *Targeted Killings in International Law*. Oxford University Press Inc.

Miller, G. (2011). Under Obama, an emerging global apparatus for drone killing. *The Washington Post*. https://www.washingtonpost.com/national/national-security/under-obama-an-emerging-global-apparatus-for-drone-killing/2011/12/13/gIQANPdILP_story.html

Miller, G., DeYoung, K., Whitlock, C., & Tate, J. (2012). *Plan for hunting terrorists signals U.S. intends to keep adding names to kill lists*. The Washington Post. https://www.washingtonpost.com/world/national-security/plan-for-hunting-terrorists-signals-us-intends-to-keep-adding-names-to-kill-lists/2012/10/23/4789b2ae-18b3-11e2-a55c-39408fbe6a4b_story.html

Minsky, M. L. (1967). *Computation: Finite and Infinite Machines*. Prentice-Hall.

Minsky, M. (1986). *The Society of Mind*. Simon and Schuster.

Minsky, M. (2006). *The Emotion Machine*. Simon & Schuster.

Mintz, A., & DeRouen, K. Jr. (2010). *Understanding foreign policy decision making*. Cambridge University Press. doi:10.1017/CBO9780511757761

Mir, A. (2018). *The U.S. drone war in Pakistan revisited*. Stanford, Freeman Spogli Institute for International Studies. https://fsi.stanford.edu/news/us-drone-war-pakistan-revisited

Miranda, G. (2008). *Simulatore di interfaccia uomo-macchina per il controllo di UAV (PhD)*. Alma Mater Studiorum, Università di Bologna., doi:10.1016/S1479-3601(2006)7

Miranda, G. (2008). *Simulatore di interfaccia uomo-macchina per il controllo di UAV*. PhD, Alma Mater Studiorum, Università di Bologna.

MSSI. (2002). *Strategic Intelligence Considerations When Preparing for War* (MSSI Thesis). National Intelligence University, Defense Intelligence Agency, Bolling Air Force Base, Washington, DC.

Muellj, D. (2010). *Redeeming Economics*. ISI Books.

Munro, C. A. O. (2014). Mapping the vertical battlespace: towards a legal cartography of aerial sovereignty. *London Review of International Law, 2*(2), 233-261. . doi:10.1093/lril/lru008

Nakashima, E., & Whitlock, C. (2011). *With Air Force's Gorgon Drone 'we can see everything'*. The Washington Post. https://www.washingtonpost.com/wp-dyn/content/article/2011/01/01/AR2011010102690.html?sid=ST2011012204147

Netti, E. (2020). *Droni, il mercato prende quota*. Il Sole 24 ORE. https://www.ilsole24ore.com/art/droni-mercato-prende-quota-ACQSyRIB

Newcastle University Official Website. (2019). *Ground-breaking conflict simulator helps train budding nurses*. Retrieved from https://www.newcastle.edu.au/newsroom/featured/ground-breaking-conflict-simulator-helps-train-budding-nurses

Niles, R. (2002). *The Strategic Intelligence of Trust Life, an Evolutionary Force of Nature*. CreateSpace Independent Publishing.

Norrman, A., & Jansson, U. (2004). Ericsson's proactive supply chain risk management approach after a serious sub-supplier accident. *International Journal of Physical Distribution & Logistics Management, 34*(5), 434–456. doi:10.1108/09600030410545463

Novembre, F. F. (2017). *Mondo Digitale: Intervista a Eugene Goostman – Divagazioni sulla Intelligenza Artificiale (Interview to Eugene Goostman – Digressions on Atificial Intelligence)*. AICA Associazione Italiana Informatica e Calcolo Automatico.

Nurmala, N., de Leeuw, S., & Dullaert, W. (2017). Humanitarian-business partnerships in managing humanitarian logistics. *International Journal of Supply Chain Management, 22*, 82-94.

Nvidia HoloDeck Official Website. (2021). Retrieved from https://www.nvidia.com/en-us/design-visualization/technologies/holodeck/

O'Connell, M. E. (2010). The International Law of Drones. *American Society of International Law, 14*(37). https://www.asil.org/insights/volume/14/issue/37/international-law-drones

O'Mara, S. (2018). The captive brain: Torture and the neuroscience of humane interrogation. *QJM: Monthly Journal of the Association of Physicians, 111*(2), 73–78. doi:10.1093/qjmed/hcx252

Oculus Business. (2020). *Building empathy to enhance hospitality*. Retrieved from https://business.oculus.com/case-studies/hilton/

Office of the Secretary of Defense of the United States of America (2005). *Unmanned Aircraft Systems Roadmap 2005-2030*. Author.

Oglesby, R., & Burke, J. (2012). *Platforms for private sector-humanitarian collaboration*. King's College London.

On Artificial Intelligence – A European approach to excellence and trust. (n.d.). White paper. https://ec.europa.eu/info/sites/info/files/commission-white-paper-artificial-intelligence-feb2020_en.pdf

Orban, V. (2014). I droni nelle operazioni di intelligence e anti-terrorismo in Yemen e Pakistan. *Osservatorio di Politica Internazionale (BloGlobal – Lo sguardo sul mondo), 2*, 1-12.

Orban, V. (2015). *Il mercato globale dei droni: profili militari e prospettive geopolitiche*, in *Intelligenza artificiale nei nostri cieli – Neuron, il drone europeo*. NoGeoingegneria. https://www.nogeoingegneria.com/tecnologie/spazio/intelligenza-artificiale-nei-nostri-cieli-neuron-il-drone-europeo/

Orozobekova, C. (2015). The US Drone Policy Under International Law. *Spotlight on Regional Affairs, 34*(11-12). http://www.irs.org.pk/spotlight/spnd2015.pdf

Outley, C., Bowen, S., & Pinckney, H. (2020). Laughing while black: Resistance, coping and the use of humour as a pandemic pastime among blacks. *Leisure Sciences*. Advance online publication. doi:10.1080/01490400.2020.1774449

Pešek, P. (2016). *The Role of Drones in Contemporary Warfare*. Metropolitan University Prague, International Relations and European Studies.

Poole, A. (2020). Re-theorising the funds of identity concept from the perspective of subjectivity. *Culture and Psychology, 26*(3), 401–416. doi:10.1177/1354067X19839070

Priest, D., & Arkin, W. (2011). *Top Secret America: The Rise of the New American Security State*. Little, Brown and Company.

Purkiss, J., & Serle, J. (2017). *Obama's covert drone war in numbers: ten times more strikes than Bush*. The Bureau of Investigative Journalism. https://www.thebureauinvestigates.com/stories/2017-01-17/obamas-covert-drone-war-in-numbers-ten-times-more-strikes-than-bush

Putman, P. G. (2012). *Virtual Simulation in Leadership Development Training: the Impact of Learning Styles and Conflict Management Tactics on Adult Learner Performance*. ETD Archive. Retrieved from https://engagedscholarship.csuohio.edu/etdarchive/249

Radsan, A. J., & Murphy, R. (2011). Measure Twice, Shoot Once: Higher Care for CIA-Targeted Killing. *University of Illinois Law Review*, •••, 1205.

Radsan, A. J., & Murphy, R. (2011). Measure Twice, Shoot Once: Higher Care for CIA-Targeted Killing. *University of Illinois Law Review*, 1201–1242.

Ravindran, A. R., & Warsing, D. P. J. (2013). *Supply chain engineering: Models and applications*. CRC Press.

Read, R. (2020). *World's most feared drone: CIA's MQ-9 Reaper killed Soleimani*. Washington Examiner. https://www.washingtonexaminer.com/policy/defense-national-security/worlds-most-feared-drone-cias-mq-9-reaper-killed-soleimani

ReportD. (2020). https://wearesocial.com/digital-2020

Riley, T. (2019). *Sea Level Rise Extrapolated from current NASA readings*. NASA. ftp://podaac.jpl.nasa.gov/allData/merged_alt/L2/TP_J1_OSTM/global_mean_sea_level/GMSL_TPJAOS_4.2_199209_201803.txt

Riley, T. (2020a). *Born to Storms*. Amazon.com.

Riley, T. (2020b). *Born to Storms Web Page*. Retrieved from website at https://bigmoondig.com/Stories/BMDStoriesBtS.html

Ripple, W. J., & Wolf, C. (2019). *World Scientists' Warning of a Climate Emergency*. Retrieved from https://academic.oup.com/bioscience/advance-article/doi/10.1093/biosci/biz088/5610806

Rogoff, B. (2003). *The cultural nature of human development*. Oxford university press.

Ross, A. S., & Rivers, D. J. (2017). Digital cultures of political participation: Internet memes and the discursive delegitimisation of the 2016 US Presidential candidates. *Discourse. Context & Media*, *16*, 1–11. doi:10.1016/j.dcm.2017.01.001

Rukeyser, M. (2018). *Science Fiction as the Mythology of the Future. The universe is made of stories, not of atoms*. https://www.centerforfutureconsciousness.com/pdf_files/Readings/ScienceFictionLongArticle.pdf

Russel, S. J., & Norvig, P. (2015). *Artificial Intelligence, A modern Approach, Third Addition*. Person Education.

Sabbagh, D. (2019). *Killer drones: how many are there and who do they target?* The Guardian. https://www.theguardian.com/news/2019/nov/18/killer-drones-how-many-uav-predator-reaper

Sauer, F., & Schörnig, N. (2012). Killer drones: The 'silver bullet' of democratic warfare? *Sage Journals, Security Dialogue, 43*(4), 363-380. https://journals.sagepub.com/doi/full/10.1177/0967010612450207

Savory, A. (2013, February). *How to green the world's deserts and reverse climate change*. Retrieved from TED video at https://www.ted.com/talks/allan_savory_how_to_green_the_world_s_deserts_and_reverse_climate_change?language=en

Savory, A., & Butterfield, J. (2016). *Holistic Management, A commonsense revolution to restore our environment*. Island Press.

Schwartz, P. (1991). The Art of The Long View. New York: Doubleday Ed.

Seattle PI. (2004). 'Master/slave' named most politically incorrect term. *Seattle PI.*

Senge, P. (1990). *The fifth discipline*. Doubleday.

Serle, J., & Sargand, P. (2015). *Afghanistan: Reported US air and drone strikes 2015*. The Bureau of Investigative Journalism. https://www.thebureauinvestigates.com/drone-war/data/get-the-data-a-list-of-us-air-and-drone-strikes-afghanistan-2015

Sfard, A. (2019). Learning, discursive fault lines, and dialogic engagement. In N. Mercer, R. Wegerif, & L. Major (Eds.), *The Routledge International Handbook of Research on Dialogic Education*. doi:10.4324/9780429441677-9

Shachtman, N. (2009). *CIA Chef: Drones 'Only Game in Town' for Stopping Al Qaeda*. Wired. https://www.wired.com/2009/05/cia-chief-drones-only-game-in-town-for-stopping-al-qaeda/

Shaw, I. G. R. (2014), *The Rise of the Predator Empire: Tracing the History of U.S. Drones*. Understanding Empire. https://understandingempire.wordpress.com/2-0-a-brief-history-of-u-s-drones/

Shaw, I., & Akhter, M. (2014, April 4). The dronification of state violence. *Critical Asian Studies*, *46*(2), 211–234. doi:10.1080/14672715.2014.898452

Shepherd, D. A., & Williams, T. A. (2014). Local venturing as compassion organizing in the aftermath of a natural disasters: The role of localness and community in reducing suffering. *Journal of Management Studies*, *51*(6), 952–994. doi:10.1111/joms.12084

Shewan, D. (2020). *How to Do a SWOT Analysis for Your Small Business (with Examples)*. The WordStream Blog. https://www.wordstream.com/blog/ws/2017/12/20/swot-analysis

Shifman, L. (2014). *Memes in digital culture*. The MIT Press.

Singer, T., & Lamm, C. (2009). The social neuroscience of empathy. *Annals of the New York Academy of Sciences*, *1156*(1), 81–96. doi:10.1111/j.1749-6632.2009.04418.x PMID:19338504

Smith, R. (2012). The utility of force: The art of war in the modern world. Penguin UK.

Somma, I. (2020). *Esperienza d'uso vs usabilità*. Telemaco. https://www.telemaco.it/blog/esperienza-di-uso-vs-usabilita/

Spatial Official Website. (2016). Retrieved from https://spatial.io/

Stadler, L., & Van Wassenhove, L. N. (2012). The Logistics Emergency Teams: Pioneering a new partnership model. Case Study, INSEAD, European Case Clearing House, ECCH (Ref. No. 712-033-1).

Sterman, J. (2000). *Business dynamics: Systems thinking and modeling for a complex world*. Irwin McGraw-Hill.

Sterman, J. (2001, Summer). System Dynamics Modeling: Tools for learning in a complex world. *California Management Review*, *43*(4).

Sterman, J. (2002, Winter). All models are wrong: Reflections on becoming a systems scientist. *System Dynamics Review*, *18*(4), 501–531. doi:10.1002dr.261

Stewart, P. (2016). *Exclusive: Obama approves broader role for U.S. forces in Afghanistan.* Reauters. https://uk.reuters.com/article/us-usa-afghanistan-obama-exclusive-idUKKCN0YW055

Tang, C. S. (2006). Perspectives in supply chain risk management. *International Journal of Production Economics*, *103*(2), 451–488. doi:10.1016/j.ijpe.2005.12.006

The Free Dictionary. (n.d.). https://www.thefreedictionary.com/unmanned+aerial+vehicle

The White House. (2009). *Promise kept.* The White House, President Barack Obama. https://obamawhitehouse.archives.gov/iraq

Theobald, O. (2017). *Machine Learning for Absolute Beginners.* Second Addition. Theobald.

Thomas, A. (2003). *Humanitarian Logistics: Enabling Disaster Response.* White paper, Fritz Institute, San Francisco, CA.

Thomas, A., & Kopczak, L. (2005). *From logistics to supply chain management: the path forward in the humanitarian sector.* Fritz Institute.

Tilcsik, A., & Marquis, C. (2013). Punctuated generosity: How mega-events and natural disasters affect corporate philanthropy in U.S. communities. *Administrative Science Quarterly*, *58*(1), 111–148. doi:10.1177/0001839213475800

Tillman, N. (2018). *The Big Melt.* South Branch Press.

Tobia, S. (2015). Close encounters: The role of culture in interrogation and questioning in the second world war. *Rúbrica Contemporánea*, *4*(8), 69. doi:10.5565/rev/rubrica.94

Tognocchi, M. (2018). *Drone warfare and the metamorphosis of battlefield: security, space and technology.* Leiden University.

Tomasini, R. M., & Van Wassenhove, L. N. (2004). *A framework to unravel, prioritize and coordinate vulnerability and complexity factors affecting a humanitarian response operation.* INSEAD working paper, No. 20 04/41/TM, INSEAD Fontainebleau, France.

Tomasini, R. M., & Van Wassenhove, L. N. (2009). From preparedness to partnerships: Case study research on humanitarian logistics. *International Transactions in Operational Research*, *16*(5), 549–559. doi:10.1111/j.1475-3995.2009.00697.x

Treccani online encyclopaedia. (n.d.a). *Interfaccia.* https://www.treccani.it/enciclopedia/interfaccia/

Treccani online encyclopaedia. (n.d.b). *SWOT, analisi.* https://www.treccani.it/enciclopedia/analisi-swot_%28Dizionario-di-Economia-e-Finanza%29/

Trunick, P.A. (2005). Special report: delivering relief to tsunami victims. *Logistics Today*, *46*(2), 1-3.

Tsing, A. L. (2015). *The mushroom at the end of the world: On the possibility of life in capitalist ruins*. Princeton University Press.

UNDRR United Nations office for Disaster Risk Reduction. (2019). *Global assessment report on disaster risk reduction.* Available at https://www.undrr.org/publication/undrr-annual-report-2019

United Nations General Assembly. (1974). *Resolution 3314 (XXIX), Definition of Aggression.* University of Minnesota. Human Rights Library.http://hrlibrary.umn.edu/instree/GAres3314.html#:~:text=Aggression%20is%20the%20use%20of,set%20out%20in%20this%20Definition

United Nations Sustainable Development Goals. (n.d.). *Transforming our World: The 2030 Agenda for Sustainable Development*. United Nations. Available at https://www.un.org/sustainabledevelopment/

United Nations. (1945). *Statute of the United Nations*. https://www.un.org/en/sections/un-charter/chapter-i/index.html

United Nations. Glossary. (1992). Internationally agreed glossary of basic terms related to disaster management. U.N. International Decade for Natural Disaster Reduction.

Vainauskas, A. (2014). *The Strategic Costs and Consequences of the U.S. Drone Warfare*. https://www.academia.edu/7532477/The_Strategic_Costs_and_Consequences_of_the_U.S._Drone_Warfare_DRAFT_

Valitis, D. (n.d.). *The Current & Future Use of Armed Drones: Analysing the Controversy*. https://www.academia.edu/11476850/The_Current_and_Future_Use_of_Armed_Drones_Analysing_the_Controversy

Van Wassenhove, L. N. (2006). Humanitarian aid logistics: Supply chain management in high gear. *The Journal of the Operational Research Society*, *57*(5), 475–489. doi:10.1057/palgrave.jors.2602125

Varanasi, R. P. (2018). *Enterprise Agility in Today's Era of Complexity. In Improving Business Performance Through Effective Managerial Training Initiatives*. IGI Global International Publisher.

Varanasi, R. P. (2019). *Future of Jobs and Humans in a Tech-Centric World. In Forecasting and Managing Risk in the Health and Safety Sectors*. IGI Global International Publisher.

VogelR. J. (2011). Drone Warfare and the Law of Armed Conflict. *Denver Journal of International Law and Policy, 39*(1), 101-138. https://ssrn.com/abstract=1759562

von Clausewitz, C. (1832). *On War* (M. Howard & P. Paret, Trans.). Oxford University Press.

Wadhams, P. (2017). *A Farewell to Ice, A report from the Arctic*. Oxford Press.

Wagner, G., & Weitzman, M. L. (2015). *Climate Shock, The economic consequences of a hotter planet*. Princeton University Press. doi:10.1515/9781400865475

Walch, S. (2014). *US Drone operations against Al Qaeda*. Oxford Brookes University.

Wang, P. (2016). *From Computing to Computational Thinking*. Chapman and Hall/CRC.

Warrell, H. (2020). *From Desert Storm to Soleimani: how US drone warfare has evolved*. Financial Times. https://www.ft.com/content/6346dd78-322d-11ea-9703-eea0cae3f0de

Warren, K. (2003). *The Critical Path: Building strategic performance over time*. Volapress.

Weiner, T. (2005). *New Model Army Soldier Rolls Closer to Battle*. The New York Times. https://www.nytimes.com/2005/02/16/technology/new-model-army-soldierrolls-closer-to-battle.html?auth=login-facebook&searchResultPosition=1

Wenger-Traynor, E., Fenton-O'Creevy, H. S., Kubiak, C., & Wenger-Traynor, B. (2015). *Learning in Landscapes of Practice: Boundaries, Identity and Knowledgeability*. Practice-Based Learning.

West, N. (2014). *Historical dictionary of british intelligence*. Scarecrow Press.

Williams, K. W. (2004). *A Summary of Unmanned Aircraft Accident/Incident Data: Human Factors Implications*. Civil Aerospace Medical Institute, Federal Aviation Administration, Oklahoma City, OK.

Wilson Center. (2012). *The Efficacy and Ethics of U.S. Counterterrorism Strategy – Transcript of Remarks by John O. Brennan*. Wilson Center. https://www.wilsoncenter.org/event/the-efficacy-and-ethics-us-counterterrorism-strategy

Wilson, E. O. (2012). *The Social Conquest of Earth*. Liveright Publishing.

Wilson, E. O. (2017). *Half Earth. Our Planet's Fight for Life*. Liveright Publishing.

Wired. (2020). Tech Confronts Its Use of the Labels 'Master' and 'Slave'. *Wired.*

Woods, C. (2011). *US claims of "no civilian deaths" are untrue*. The Bureau of Investigative Journalism. https://www.thebureauinvestigates.com/stories/2011-07-18/us-claims-of-no-civilian-deaths-are-untrue

Woods, C., & Ross, A. (2012). *Revealed: US and Britain launched 1,200 drone strikes in recent wars*. The Bureau of Investigative Journalism. https://www.thebureauinvestigates.com/stories/2012-12-04/revealed-us-and-britain-launched-1-200-drone-strikes-in-recent-wars

World Bank. (n.d.). Retrieved from: http://www.worldbank.org

World Meteorological Organization. (2019). *The Global Climate in 2015-2019*. Retrieved from https://library.wmo.int/doc_num.php?explnum_id=9936

WTO – World Trade Organization. (n.d.). Retrieved from: http://www.wto.org

Xiao, H., Zhang, Y., Kong, D., Li, S., & Yang, N. (2020). The Effects of Social Support on Sleep Quality of Medical Staff Treating Patients with Coronavirus Disease 2019 (COVID-19) in January and February 2020 in China. *Medical Science Monitor, 26*. Advance online publication. doi:10.12659/MSM.923549 PMID:32132521

Ye, L., & Abe, M. (2012). The impacts of natural disasters on global supply chains. Working Paper Series No. 115, June 2012 in Asia-Pacific Research and Training Network on Trade (ARTNeT), Bangkok, ESCAP.

Zambernardi, L. (2010, July). Counterinsurgency's Impossible Trilemma. *The Washington Quarterly*, *33*(3), 21–34. doi:10.1080/0163660X.2010.492722

Zeleny, J. (2007). *As Candidate, Obama Carves Antiwar Stance*. The New York Times. https://www.nytimes.com/2007/02/26/us/politics/26obama.html

Zenko, M. (2011). *Who can't America kill?* Council on Foreign Relations. https://www.cfr.org/blog/who-cant-america-kill

Zenko, M. (2012a). *How many civilians are killed by U.S. drones?* Council on Foreign Relations. https://www.cfr.org/blog/how-many-civilians-are-killed-us-drones

Zenko, M. (2012b). *Targeted Killings and Signature Strikes*. Council on Foreign Relations. https://www.cfr.org/blog/targeted-killings-and-signature-strikes

Zenko, M. (2012c). *Daniel Klaidman's Revelations*. Council on Foreign Relations. https://www.cfr.org/blog/daniel-klaidmans-revelations

Zhabotynska, S. (2020). The narrative multimedia concept: An algorithm for the analysis (internet-memes about COVID-19). *Cognition, Communication, Discourse*, *20*, 92–117.

About the Contributors

Luisa dall'Acqua is a senior cognitive scientist. She is an author, reviewer, and editor for several international scientific journals and book editions. Since 1990 she performed teaching activities, training students, teachers in service, headmasters, and professionals; and since 2003 she was a member of international projects and research teams (USA, Europe, Asia). After a MSc in Philosophy, Luisa received a Ph.D. in Sociology of Legal Institutions and Policies, a second Ph.D. in Psychological and Social Sciences, and a Specialization Master in Risk and Disaster Management. The main field of research is an engineering and social-cognitive approaches to intelligence and problem solving, and in particular to the decision-making risk under stress, conflict, and unpredictability, ranging from individual, collective, institutional intelligence applications to computational intelligence. Currently, she is a permanent teacher at the TCO Scientific Lyceum (Italy) and an adjunct professor at the University of Bologna.

* * *

Pedro B. Água is a Professor of General Management at the Portuguese Naval Academy. He has authored several articles and book chapters, while continuing his research in the field of cutting-edge technology, industrialization, innovation and business policy and regularly teaches such. Professor Água has over twenty-five years of experience across high technology endeavours, from defence to telecommunications and oil and gas industry, combining his extensive professional and business background with teaching. Andre holds a Ph.D. in Management and Engineering awarded by the University of Lisbon.

Immacolata Carcarino is a former Italian Air Force Warrant Officer with eighteen years of service. During her military career, Immacolata specialized in Force Protection and Personnel Recovery. She is a qualified NATO Lessons Learned Staff Officer and was responsible for managing lessons learned processes in various international activities, including operations, training events and meetings. After a Bachelor's degree in Languages and Cultures at the University of Rome "La Sapieza", Immacolata graduated in International relations and diplomatic affairs at the University of Bologna.

Anacleto Correia is a professor of Information Systems at the Portuguese Navy Academy. He holds a Ph.D. in Computer Science, a M.Sc. in Statistics and Information Management, a B.Sc. degree in Management, and also a B.Sc. in Naval Sciences from the Portuguese Naval Academy. His research interests are focused on requirements and software engineering, process modelling, data mining, machine learning, and business engineering. Professor Correia has more than 15 years of industry experience, leading projects and architecting large scale software projects, and is the author of several scientific papers in journals and conference proceedings.

Armindo Frias is a Military Professor of Logistics, Management and Administration at the Portuguese Naval Academy. He holds a PhD in Management, a MSc in Business Sciences, a BSc in Naval Military Sciences - branch of Naval Administration. He has more than thirty years of experience in Naval Logistics, Public Administration and Budgeting. Armindo Frias is the author of several scientific papers in journals and conference proceedings, his main research interests are naval and maritime logistics, integrated logistics support, systems engineering and decision making.

Irene Gironacci is currently working as Project Manager on Extended Reality research projects at Swinburne University of Technology (Australia), and specializing with a PhD in Extended Reality and Artificial Intelligence. Previously, she has been working as Mixed Reality Engineer on a H2020 project for training security agents at Luxembourg Institute of Science and Technology, Luxembourg. Her research focuses on the following topics: new technologies, project management, risk management, artificial intelligence, 3D graphics, motion capture, cinematography.

Maria Antonietta Impedovo, Ph.D., is Associate Professor at ADEF Laboratory, Aix-Marseille University, France. She teaches at the School of Education at Aix-Marseille University, France. She held her Ph.D. about Educational technology in Italy, followed by a postdoc in Swiss. Her main research interests are identity, teacher agency, and teacher professional development in a sociocultural psychology perspective. Since 2010, she has been actively participating in multiple national and international formative and research projects (like the capacity building "Blended learning for teacher educators between Asia and Europe". She has written scientific papers in Italian, Spanish, French, and English. Last book (2021): Identity and Teacher Professional Development: A Reflective, Collaborative and Agentive Learning Journey (Springer Briefs in Education).

Maria Luisa Nardi is a Historian of Political and Economic Institutions. She performed as a high school teacher for more than 30 years, and during this time she was awarded for innovation and experimentation in teaching methods and forefront technology. She conducted research in Economic History, starting from Greek and Roman times to the current context. Her work offers valuable empirical and theoretical insights into the development of market-supporting institutions, their long-term consequences, and their relationship to state-building. Her historical review discusses recent advances in understanding the precise Mechanisms responsible for the persistent effects of institutions on economic development.

Giulia Casmiro Scarmagnani is a recent graduate in Diplomatic and International Sciences. After studying International Relations at the University of Trento, she decided to pursue the specific branch of international security, studying for the master's degree of the same name at the Alma Mater Studiorum University of Bologna. During this last cycle of studies, she became interested in the subject of Remotely Piloted Aircraft, collaborating with esteemed personalities from both the international and engineering worlds. She is currently working as a corporate executive assistant, while cultivating her passion for intelligence and security, both at the national and international level.

Bobby Varanasi is one of the acknowledged Top 25 Globalization Leaders in the global sourcing space and the Founder of Matryzel Consulting – an independent New York headquartered advisory firm that has been acknowledged as one of the World's Top 20 "Best Outsourcing Advisory Firms" three years in a row (2013, 2014 & 2015). He brings with him two decades of experience in consulting and management across Technology, Business Services and building global operations. He advises federal governments on ICT sector development with particular emphasis on policy transformation, and industry-government partnerships aimed at creating domestic resilience through increasing productive (and inclusive) faculties of entrepreneurs. Bobby also advises Fortune 500 customer organizations and emerging market entrepreneurs on strategy, growth, sourcing and expansions. Bobby is the author of a book titled "Humanomics – Making Sense of Socio-economic Impacts of Global Sourcing". Bobby holds a Diploma in Sustainable Development from Columbia University, USA; and two Masters Degrees (one each in Solid State Physics and Management) from Berhampur University, India.

Anna Verrini graduated in Mathematics at University in Torino. At the beginning she worked for a long period as a computer specialist, being responsible for the development of many projects in very different areas. At the moment she is involved, as independent consultant, with some team in activities referring the ehealth applications, AI and the Computer Ethics and the professional profiles analysis.

Francesco Saverio Zamboni has a master's degree in International and Diplomatic Sciences, curriculum Politics and International Security, at Alma Mater Studiorum University of Bologna, with a final dissertation in Intelligence and Political Decision Making. Previously, he earned a bachelor's degree in Political Science and International Relations at the University of Roma Tre. Passionate about geopolitics, he focuses his studies on the relationship between politics and security with a specific focus on economic and securitarian issues. Since January 2021, the collaboration with the think tank, Mondo Internazionale, is articulated in this direction where he currently holds the position of senior researcher in the economic division of the Geostrategic Earth Observations team.

Index

H

I

K

M

N

O

P

R

S

U

V

www.ingramcontent.com/pod-product-compliance
Lightning Source LLC
LaVergne TN
LVHW061542290926
849162LV00004B/882